Writing Legal Descriptions

IN CONJUNCTION WITH
SURVEY BOUNDARY CONTROL

by

GURDON H. WATTLES

Library of Congress Card
Nos. 75-36431 & 78-68650

Published by:
Wattles Publications
P.O. Box 1706
Tustin, California 92681-1706

987654

Dedication

To those individuals who write legal descriptions is this book dedicated. May all who read it find therein a valuable tool to be used in rendering a much needed service to the profession.

Introduction

The simple title selected by Mr. Gurdon H. Wattles for this book indicates at its best the attitude that should prevail toward one of the more important aspects of modern land surveying. As indicated by the author, it is not the intent of this work to be an exhaustive treatise on the legal background involved in removing discrepancies in locating land. Rather, it is a workable set of fundamentals, critically defined, and set forth with the necessary relationships to insure a certain degree of rigor.

Before proceeding, an introduction to the pertinent history necessary for suitable background is shown. Instances where this has been reiterated and confirmed by precedent cases in the courts are presented. Full note is made of code and other statutory definitions, both Federal and State, which are today applicable to the problems of surveying, mapping, description, and conveyancing.

The concept of land surveying as an art is subtly covered throughout the text. In effect, the modern surveyor is charged with making precise measurements using a rapidly changing available technology. To a certain degree, these are discussed in certain types of definitions which have slowly become part of standard nomenclature of land survey work. However, it is not the problem of defining the techniques of making these skilled measurements to which Mr. Wattles addresses himself; rather, it is the more important question of showing the situations that arise when the surveyor, in the course of performing these measurements, finds that he is unable to reconcile those implied by the record with those disclosed by evidence in the field. It is at this juncture that the surveyor is called upon to exercise his skill in the art of land surveying. It is emphasized that there are no particular formulae that encompass a solution to all problems, but methods of considering alternatives of what is implied by the apparent discrepancies in the field evidence versus the written record are set forth. At all times the classic precept of following in the footsteps of the original surveyor receives its due.

To avoid future discrepancies and to improve the conditions under which the surveyor of the future might work, the concept of locatability of a description is heavily emphasized. This is of paramount importance in the attempt to minimize ambiguity and reduce discrepancies. Suitable examples from decisions rendered by the courts holding heavily in favor of possibility of location that can be readily laid out in the field are presented. Since modern land development is occurring in areas previously thought to be of little value, the traditional two dimensional frame work of the land surveyor is dropped. Instead, full realization is made that the modern surveyor is called upon to work within the real world of land, sea, and air.

The relationship of description and surveying problems that arise on land having water boundaries receives a modern update. The interest in shoreline resources is most evident today, and the need for the skill of the surveyor in handling location problems affecting lands in this vicinity cannot be over estimated. Similarly, because of the requirements of mankind for space, modern development has forced consideration of air space for use as possible physical building sites. This can be, of course, above ground with the concomitant conditions of light, air, or unobstructed view being given consideration, or it might be substantially below the ground surface with appropriate attention paid to interpretation of utilities, drainage conditions, and features affecting the stability of foundation.

As tools to the writing of descriptions, a good review is made of the concept of what constitutes a record — public, private, and quasi-public—and the sources of same. These can be in original form or stored in some second or third generation form involving a more sophisticated technology. Retrieval of this information from the simple hand written record or data banks is reviewed. Also considered are adjuncts to this process, the concept of collective references to records which can satisfy simple disclosure or constructive notice tests and those of use for the technical side of surveying such as coordinate systems, both general and local.

The widespread variance in the means of defining orientation is discussed; in particular, the need for the description writer to be aware of some of the more technical differences in types of orientation are presented.

Accompanying each chapter are suitable problems which emphasize the points of discussion. Some of these are answered in brief form because of singularity of the answer; others are discussed more fully where such solutions are warranted.

The Table of Contents indicates the classic overview of large topics such as control, boundaries, analysis and interpretation, monuments, the general subjects down to the more refined details of items such as reversion of title resulting from vacation or abandonment procedures.

The author is most qualified to expound on the subject matter contained therein, with many years in a position of responsibility for a large land title organization. His activities are many fold—teaching, construction, land planning, civil engineering, land surveying, and other aspects of land development. Most important of all, his activities with his peer groups— the professional engineer and surveying organizations for which he has given much service as officer, and committee member and chairman—have accorded him recognition at the national as well as regional level. As a frequent lecturer before many technical groups he has always emphasized the need for inspecting the evidence for consideration of the alternatives implied, and above all he has set forth the cautionary precept to avoid looking for the easy solution to wrap up the problem at hand. As always, he has shown his respect for the famous adage of another Wattles: — the possibility that the contrary may be shown.

To the practicing land surveyor, civil engineer, title man, appraiser, attorney, land planner, and all those concerned with the orderly development of our land, I recommend this work without hesitation. Unlike the noted New England surveyor, Henry Thoreau, this concern cannot be content with just doing enough for today—provision for tomorrow must be a rational reality.

Ira H. Alexander
Assistant Chief Deputy
Department of County Engineer
County of Los Angeles
Los Angeles, California

January, 1976

Preface

Because land surveying uses legal descriptions as a directive and because descriptions are most always based upon and developed from survey field work, the two are inevitably intertwined. For this reason, knowledge in one part without some comprehension of the other makes for instability in the final product.

It has been the experience of many to read legal descriptions for the purpose of making surveys and/or drawing maps from them — and occasionally with confused comprehension because of the wordings used. Sometimes these same individuals are asked to create new descriptions and when they attempt to do so, find the approach to writing is considerably different from their experience in reading them.

Recognizing these confrontations, the objective of this book is to provide some guidelines for writing descriptions of land. Although "legal" description is generally the nomenclature used, the writing is specifically directed to and for the purpose of delineating a certain area of land or space which cannot apply to such a condition anywhere else. The adjunct of the word "legal" pertains to the fact that the description must be able to withstand attack under law.

For these reasons, several disciplines of knowledge must be considered and a working comprehension of them acquired in order to develop the ability to properly compose a legal description.

The citations of court cases, federal and state codes and examples of survey problems are the result of choosing pertinent ones from extensive research and are given as a part of understanding the issues involved. From this, the correct procedure for writing descriptions unfolds step by step.

This book is in fact a compilation and expansion of the material developed and used by the author and distributed in his classes and seminars conducted over many years on this subject matter.

It is recognized that hundreds of examples could be cited and hundreds more court cases could be mentioned but it is the objective of this textbook to hit the highlights and cover fundamentals. Furthermore, the professor using this book without doubt has many examples and court cases that apply to similar situations in his locale which can be woven into the discussion at the appropriate places.

The development of a notebook to be had at your fingertips containing copies of those state and federal government statutes, code sections, rules and court cases particularly applicable to the area wherein you are working will prove to be a valuable adjunct.

It is assumed that the reader is already acquainted with some of the subject matter within certain areas of study here; therefore, time and space is not devoted to reviewing much material that is found in published works.

The author wishes to express not only his appreciation but also his gratitude for the assistance received in the preparation of this book, namely: Mr. Ira Alexander, Assistant Chief Deputy County Engineer, County of Los Angeles, California, a recognized authority and teacher on surveying, for his review and constructive suggestions of the manuscript; Mr. John R. Y. Lindley, Associate Counsel, Senior, for Title Insurance and Trust Company for his meticulous examination of the legal aspects of this dissertation; Mr. Thomas E. McKnight, Vice President of Title Insurance and Trust Company, a nationally recognized specialist on water boundaries for his inspiration and help over the years and particularly his review and aid on Chapter 6, Water Boundaries; and to the many individuals in Title Insurance and Trust Company and in Pioneer National Title Insurance Company whose assistance and moral support has always been available and wholeheartedly given.

The Problems, Questions and Answers given herein are only an indicative guide for the professor to enlarge upon or for the self teaching student to review and expand through further research.

Gurdon H. Wattles
1976

1 2

Contents

Chapter **3** *Description Fundamentals*

Chapter **4** *Determination of What Controls*

Chapter **5** *Boundaries – General*

Chapter 5 cont'd.

Chapter 6 *Water Boundaries*

Chapter **7** *Analysis and Interpretation of Descriptions*

Chapter **8** *Monuments*

7
8
9
10
11

Chapter **11** *cont'd.*

Chapter **12** *Easements and Rights-of-Ways*

Chapter **13** *Descriptions of Section Land*

12

13

14

Appen.

Index

Chapter 1

Background and Framework

History

Centuries ago, the transfer of property was accomplished by the seller and buyer walking around the perimeter of the land and observing the monuments at each corner, known in those days as bounds, or, if a monument was not there, they would set one. Later, the process became known as feoffment, the planting of a feud or fee, accompanied by a ceremony of the seller picking up a piece of stone or twig from the land and handing it over to the buyer, thus symbolizing the consummation of the transfer. This action was also called "livery of seizin" and was common in England, but the Colonists did not establish the practice of this tradition in their new-found land.

A similar custom was used out West in the land under Spanish and Mexican rule. A subject would present to the Governor his diseño (map) and expediente (petition) representing his request for an allotment of land. When it was found satisfactory, an official paper was prepared and returned carrying certain provisions, including the requirements that a permanent building be erected and occupied within one year, the land be fenced or enclosed but reserving the rights of use of public roads and easements, any rights of previous inhabitants (Indians) be acknowledged and protected, and that the transferree obtain from the local magistrate the act of juridicial possession which would define and measure the boundaries.

When this was accomplished and confirmation was given by the magistrate, the grantee would symbolize the finality by pulling up clumps of grass, breaking branches from trees or bushes, and throwing rocks to the north, south, east and west to demonstrate his rights as an owner which could not be exercised by others.[1]

In America, the Plymouth Colony in 1640, Connecticut in 1639 and Massachusetts in 1641, required by law that a notice of the conveyance of land was to be recorded. This system of public disclosure of land transfers was copied by other areas across the country and later became a part of each state's law.

After the colonies became the United States of America, there were acquisitions of large areas of land such as the Louisiana Purchase, the Mexican Territory, the Oregon Territory, etc. Throughout the colonies, the method of describing land for its transfer was by metes and bounds. The founding fathers, however, in looking over the newly-acquired country, realized that if it were to be sold quickly in order to gain revenue which was badly needed, some method other than the one in common use would have to be developed.

Public Lands

Consequently, the Continental Congress called for research into a simpler and more uniform method of dividing land. Fundamentally, the approach went on the basis of the rectangular grouping of one mile square units. The question was primarily how many and the debate narrowed to either 6 by 6 or 10 by 10 squares to a township.

[1] from *Designs on the Land,* by Robert H. Becker.

The Land Ordinance of 1785 was enacted in session March 4 and passed May 20, 1785, and became known as "The United States Rectangular Land Surveying System" and was based on the theoretically six mile square township. As experience was gained in establishing and surveying the sections and townships, subsequent acts were passed in 1796, 1800, 1805, 1812, 1820, 1832, 1836 and 1849 to amend the ordinance and, in addition, rules were established for the better performance of survey work.

General Instructions were issued in 1815, 1831, 1834, 1843, 1846, 1850 and 1851. Each subsequent set carried improvements based on prior experience. The one directed to the SURVEYOR GENERAL OF OREGON in 1851 was used in California until the first Manual was created. Another set was issued in 1851 to those surveying in Wisconsin, Iowa, Minnesota and west of the 5th. P.M. In 1855, a Manual of INSTRUCTIONS TO THE SURVEYORS GENERAL OF PUBLIC LANDS OF THE UNITED STATES was published by order of the Commissioner of the General Land Office. Later editions became known as the MANUAL OF INSTRUCTIONS FOR THE SURVEY OF THE PUBLIC LANDS OF THE UNITED STATES and have been issued in 1881, 1890, 1894, 1902, 1909, 1919 (which in fact was an advance copy of the first six chapters of the 1930 Manual), 1930, 1947 and 1973.

The use of sections, quarter sections and lots in sections with the township and range numbers and the name of the Meridian for the total area embraced make the description of land much easier and faster. Interestingly too, this method provides a self-sufficient non-duplicating vehicle for indexing such land.

A pertinent point to remember is that the Land Ordinance required a survey to be made and approved before a patent was issued. Public Lands will be treated in more detail in Chapter 13.

To understand the different laws that govern in their respective ways, we must recognize the difference between public lands and private lands. The private lands were those colony lands held by grants from England, Spain and France, ranchos held by grants to individuals from the Spanish and Mexican Governments, and other smaller grants from such countries as Germany, Portugal, etc. Also under this heading are homesteads, mining claims, land acquired by individuals from the government under patents for tide lands, swamp and overflowed lands and section lands.

Public lands are those held by the government, both federal and state, lands acquired by treaty or purchase, not including those private grants previously made, tide land and swamp and overflowed lands not disposed of, submerged lands and offshore lands, and those not included within the vacant lands retained by the Colonial states, the other New England and Atlantic Coast states (except Florida) and later by the states of West Virginia, Kentucky, Tennessee and Texas.

States shown within the area of PUBLIC DOMAIN on the map herewith (including Alaska) are those embraced under the jurisdiction of the Bureau of Land Management, Department of the Interior of the United States of America.

Private Lands

Besides the private holdings in the Colonial states, Kentucky and Tennessee, there were other areas under private grants within the Public Domain which had to be honored and circumvented by the surveys of the public lands. Some of these were the Greeley French Claim surveys in Michigan, the French Grant in Ohio in 1795 and a second one in 1798, the Virginia Military District in Ohio and others.

Following the Treaty of Guadalupe-Hidalgo concerning the Mexican Territory acquisition, the United States agreed to honor and confirm the grants of land, known as ranchos, made previously by both the Spanish and Mexican Governments to individuals. For this purpose, the District Land Commission was established to accept and act upon the diseño and expediente submitted by the ranchero. This in turn was transmitted to the Surveyor General in the General Land Office in Washington, D.C., which assigned a Deputy Surveyor to make the boundary survey of the land, and from which a metes and bounds description was drawn. The map of the survey and the description were then presented to the ranchero for his approval. If it did not conform, as in one case where his house and garden were omitted, the survey was ordered redone and a new map and description made. Following approval, the patent was issued by the U. S. Government to confirm his title.

Texas presents a distinct history of its own. Having been under the rule of six different flags, it has derived various characteristics pertaining to legal descriptions and surveys. When she wanted to have her ungranted lands included in the U. S. Rectangular Surveying System, the Department of the Interior reportedly did not have

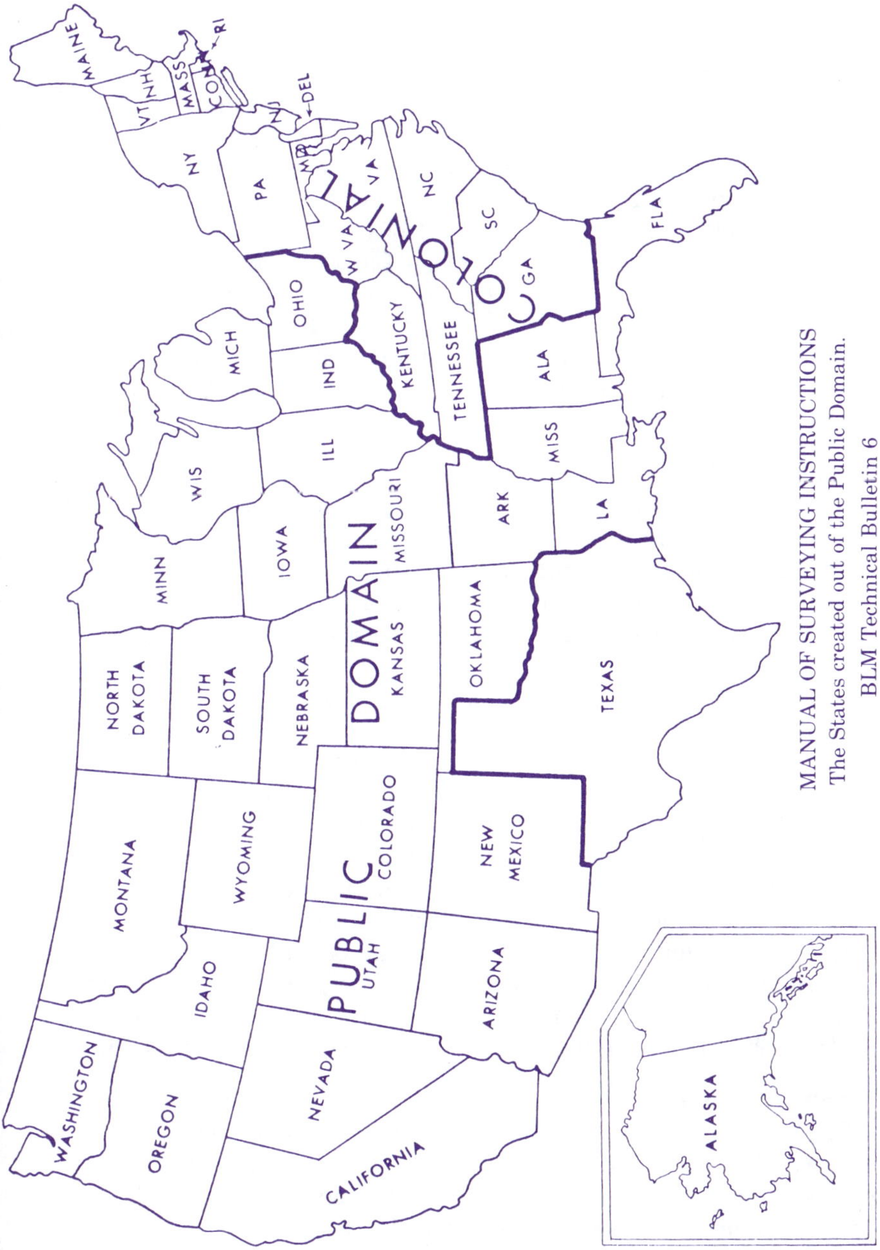

MANUAL OF SURVEYING INSTRUCTIONS
The States created out of the Public Domain.
BLM Technical Bulletin 6

enough money to provide the necessary services, so Texas had to go her own way.

Note that after government (public) land is patented to individuals, it is classified as private land, but it is still subject to its outside boundary control based on the rules for the survey of public lands.

The conclusion to be observed here is that surveys and descriptions concerned with "public lands" must observe the rules set forth in the manuals and instructions issued out of the office of the Bureau of Land Management (formerly General Land Office), Title 43 of the Federal Code and court cases which confirm, interpret or negate any of them.

Surveys and descriptions on all other lands must consider those laws and ordinances and court cases which deal with the private land aspects.

Transfers

The transfer of title from the government to an individual is by the instrument known as a patent. The conveyance of private lands between individuals, corporations or other entities, or, to or from city or county entities, is accomplished by a document called a deed. Of the latter, there are different forms such as the grant deed, quitclaim deed, corporation deed, warranty deed, statutory deed, etc. It is also possible, although not too common, to grant a parcel of land by proper wording in a certificate on a map. The reason for having a public office and a depository for deeds, etc., is so that documentary evidence is made public. There are rules concerning the procedures of depositing, recording and indexing these records so that the public can investigate them and retrieve any particular document for review.

Some instruments convey only a certain interest (a limited ownership) in real property, while others are used to encumber real property to secure loans and to make, change or add interest in and of parties to a transaction. Still other instruments and papers are made to accomplish the effect of releasing, or showing the satisfaction of, and thereby cancelling the subsequent effect of a recorded document.

Public Records

The process of recordation according to law is also known as giving constructive notice. In other words, upon inquiry into the actual notice of circumstances as to a particular fact, the person who deals with the land has constructive notice of the fact itself in all cases in which a prudent man, by prosecuting such inquiries, might have learned such fact.

Another group of items which a surveyor uses in his research of public records for information preliminary to making the field survey include also the county[2] recorder's filed maps or miscellaneous filed documents which, although usually located in the same office or building, are not officially "recorded" or indexed in the official or statutory grantor-grantee index.

Quasi Public Records

There are other sources of information known as *quasi* public records which include maps, and survey field notes found and filed in the offices of the county surveyor, county flood control, sanitation district, city engineer, state highways, water resources, power companies, railroads, etc., etc. Information from these sources often proves to be of considerable value to the surveyor in his preliminary reconnaisance to establish the framework for final boundary determination. However, let the scrivener beware of using *quasi* public records for references or control in legal descriptions because there is no law that prevents the alteration of those field notes or maps. The danger is that if a point shown in a field book or on a map and used for a control in a description is changed, there is no public notice of the change and the value of the point is lost, thus invalidating that part of the description unknowingly until other circumstances bring the faux pas to light.

Maps filed in the assessor's office and township plats filed in the district land office are also public records available to the public but not officially recorded.

Another place where a different type of quasi public record is found is in the Clerk's office of the county and/or city. The papers filed there include court cases, ordinances, resolutions and minutes of the county board or city council, etc.

The courts have held that even though many records may be classified as public, including some that are technically "quasi" public, the distinction must be made between them and those conforming to the rules set forth for constructive notice.[3] The application of this will be treated further in Chapter 11.

[2]Although the word "county" is used here, it refers also to the other political subdivisions of a similar area and type known as "parish" or "borough" where applicable.

[3]Stearns v. Title Insurance & Trust Co., 18 CA 3d. 162.

Land Registration System

The land registration system is found in history as far back as the 13th. century in Bohemia. Sir Robert Torrens of England went to Australia and developed it there in the early years where it became known as the Torrens System. Rules were established and followed, including surveying and mapping under rigid control, and Australia carries a part of her individual land ownership records under this form of registration.

In the United States, several states use land registration systems. The principle is based upon registration of a parcel of land through an application to the court by the owner, processed by an attorney who submits the legal description of the property with a map of a survey showing the boundaries and a list of abutting owners and any other interests. In proceedings required by the land registration law, every owner of property boundary to the applicant's land and every owner of any interest in said land must be notified of the hearing and given his "day in court." This, in effect, creates a quiet title form of process so that when the court decrees that the land under consideration in the application is fit to be registered, all claims to and from boundary owners and owners of instant interest have presumably been quieted or acknowledged and the Register must so show.

The depository for land registration records is usually separate from that of the public records of deed books.

Except in rare cases, such as Follette v. Pacific Light and Power Corporation, 189 Cal. 193, where there was a lack of notice which should have been given to an occupant with a claimed interest in the property, and hence a lack of fundamental due process of law, it is to be assumed that any land ownership carried in the books under this registration system has been properly processed through the court and therefore carries its full weight and retains the effect of having been through a quiet title action.

Land Data Bank Systems

Even though the old practice of making public records by copying each presented document with painstaking longhand in a musty back room of the old court house still prevails in a few places, many recorders' offices have supplanted the handwork with photographic copying but still use cumbersome books for the repository.

The next step in many areas has been the use of microfilm copy of not only the old books but of all the new documents in order to conserve space. The microfilm shots are placed in cards and called microfiche which are retrieved either manually or by automation.

Every depository must have an index system and the grantor-grantee method has been and is today the general standard used in the recorders' and the register of deeds offices. Title company plants are based on land parcel indexing instead of name indexing, although the names are listed correlatively. Approaches to diverse methods of assignment and retrieval have been developed, however, because of the demand of business and professional practitioners for faster and more direct access to information related to particular parcels of land by some descriptive reference other than through the name index.

Some of these in use in America now are: street address, block and parcel, Federal Rectangular Survey System, map based systems, Dual Independent Map Encoding (DIME), Federal Survey based coordinate system, and the State Plane Coordinate System (SPCS). A variety of other applications are being used or developed in foreign countries.

Under sponsorship of the American Bar Foundation (ABF), a conference with the theme of "Compatible Land Identifiers — The Problems, Prospects, and Payoffs" (CLIPPP) was held January 20-22, 1972 in Atlanta, Georgia† to study systems of land parcel identifiers and indexing which would meet the needs of all users of land related data, including land title records as well as ecology, land use and planning, etc.

Additional proposals made at that CLIPPP conference included: latitude-longitude designation, Mercator Grid systems, an expansion of the Federal Information Processing Standards (FIPS) coding system, a geodetic survey based system, a visual centroid applied coordinate system, a Min-Max identifier using external coordinate limits for identification, and a very sophisticated integrated land-data system using several different geographic coding structures.

Another, the North American Conference on Modernization of Land Data Systems (MOLDS) was held in Washington, D.C., April 14-17, 1975* in an effort to help further evolve land data systems which would satisfy the expanding needs of our society by making better use of advanced technology.

†Papers and discussions published in proceedings as LAND PARCEL IDENTIFIERS FOR INFORMATION SYSTEMS, by D. David Moyer and Kenneth Paul Fisher, American Bar Foundation.

*Papers and discussion published in PROCEEDINGS OF THE NORTH AMERICAN CONFERENCE ON MODERNIZATION OF LAND DATA SYSTEMS (A MULTI-PURPOSE APPROACH), North American Institute of Modernization of Land Data Systems.

It is not the purpose here to delve into the details of these diverse methods, their applicability or respective values; there is extensive reading matter on this subject available to those who wish to pursue it. Suffice it to say that insofar as legal descriptions are concerned, the land data bank system can be used to its best advantage in this area where satisfactory maps are available for the indexing reference. Where tract maps and the like are available *that have been carefully checked* (note the italics!), the reference to the lot or parcel number and the tract with proper recordation information is fine — but — in the case of metes and bounds descriptions, map reference indexing is only acceptable when, as and if the same detailed preparation has been made and very careful checking is conducted on that map. The continued surveillance of metes and bounds descriptions is a must.

Much of the foregoing is concerned with paper information. Weaving this in with facts obtained by field survey work forms the pattern from which the properly worded final description should emerge as a clear and comprehensive document.

Coordinates

Another subject which has been commanding increasing consideration is coordinates. They constantly prove their value as a tool in the fields of engineering and surveying projects.

The validity of their use in legal descriptions varies across the country. Some states accept them. Others have only established the State Plane Coordinate System for use but without giving points so designated the right of legal control. California definitely relegates coordinates to secondary value. Here again, a careful study of the codes and court cases in *your* area will reveal the status of coordinates in regard to legal descriptions. Additional discussion of this matter will be given in Chapter 11.

Laws

Besides maps, records and other miscellany of information, the framework of law both state and federal expressed in statutes and relevant court cases play a predominant part in the guiding and controlling of such matters as the filing of maps; recording of documents; surveying of land; colonial land boundaries; the public land system; the meaning of legal descriptions; land and water boundaries; monuments; agreement lines etc, etc. This fund of information is to be found in a law library and one who aspires to write descriptions that carry legal authority as well as the technicalities of applicable surveying practice will find that research time spent in such a library is worth while and rewarding.

LAW LIBRARY

PROBLEMS

1. What is feoffment?
2. What was the ceremony called whereby a transfer of ownership in real property was consummated by the seller handing a stone or twig to the buyer?
3. What act was performed by the buyer of land out west to symbolize his ownership?
4. What is the name attached to the Land Ordinance of 1785 which is concerned with the surveys of Public Lands?
5. Name 4 types of Public Lands.
6. What is the difference between public and quasi-public records? Explain.
7. What is the Land Registration System?
8. What are Land Data Bank Systems?

— NOTES —

Chapter 2

Support Information

Whether it is relative to preparation for making a survey or laying out a map, when you are reading a description, you often find many references are made to other lands for the purpose of a boundary or point of beginning and it is necessary to locate that description reference. It may be to a map or it may be to another deed on record or some other type of documentary evidence. Citations vary in the extent of information given, all the way from just a name or book and page to a complete coverage such as, "land described in deed from Willard Hoyt, a single man, to Gaylord Holmes and Arline Holmes, brother and sister, holding each an undivided one-half interest, recorded July 16, 1928 as instrument no. 3642 in Book 4951, Page 72, of Official Records in the Office of the County Recorder of said County." In the office of public records, the numbered book and its page can be easily retrieved, or, as in some areas where no book and page number is used, the instrument number on a certain date must be retrieved. If only a name is given, there are generally two avenues where you can look for the name of the owner and find a reference or a direction to a document containing a description: one is the office of the public depository of records and the other is the assessor's office.

Via Official Records

The public recorder's office will have the grantor-grantee index whereby you find opposite the name a reference to the book and page of the document and the kind of document. You then find the book and look at the description on the pages referred to.

Via Assessor's Office

The assessor's office customarily has a list of the names assessed and reference to their property; however, sometimes the descriptions used by the assessor are in a short form of their own style and usually not acceptable from a legal standpoint. Furthermore, the name assigned to that particular piece of land may not be that of the owner but only one who is paying the taxes, such as a lessee.

If you know the location of the lot in a tract, or if you have an address, it may be easier to check with the assessor's records to find the name or description of the property, but, in the end, the reference to the book and page of the deed in the public recorder's office and then the description used in that document is the proper and more reliable source of information.

Via Maps

Sheets of paper or cloth or mylar of specified dimensions which contain drafting of areas of land cut into parcels or strips with appropriate certificates and explanations concerning the land are called "maps" or "plats" depending upon the custom prevailing in the area where they are used. Any reference here to "maps" also includes "plats" where it is appropriate.

If a subdivision map is properly recorded so as to conform with the rule of constructive notice, in other words, properly indexed in the grantor-grantee index, then you may take the number of the lot in that particular tract in which you are interested and check through the grantor-grantee index and thereby obtain by the most recent conveyance the name and the book and page of the recording of the document of the present owner. Sometimes you will find help where there are private surveys filed, either in the office of the public records, or in the county surveyor's office, or sometimes, state highways maps will show the names of the owners of properties along the way, especially where they have acquired parts of those holdings for new highways. This again is not the final information upon which you will rely, but it can be helpful in your search and also, the maps themselves can give you a better picture of the situation with which you are working. In addition, railroad maps, sanitation district maps and public utility maps often include helpful information.

A variety of maps come under the heading of public records. Some states and counties and cities have many while others have fostered only a few. Maps which show a subdivision of an area of land are called subdivision maps or tract maps and, besides the delineation of lots, they also show streets, sometimes alleys, walkways and occasionally parks that are dedicated to public use. When these are recorded in the public office, they offer the media for the simplest type of description, that of merely referring to the lot and the tract according to the map recorded in the specified book and page, etc.

The condominium map is usually a one-lot subdivision and the diagramatic plans for the condominium may or may not be filed with that map. The layouts of the units are not a part of the tract map and if not filed with it, they are recorded with the declaration of restrictions.

In some areas, there is another form of subdivision map called a parcel map which is similar but does not have all of the requirements of the subdivision map.

There is often a provision for a reversion to acreage map, called a cancellation plat in some areas, which, in effect, cancels out the subdivision of a land area into many lots and reverts that area to acreage for the purpose of lower taxes or to be used in some other manner, such as a shopping center.

Some states provide for "official maps" which are for the purpose of permitting a governing body such as the city or county to file a map for clarification purposes. For example, this would be used in the case of an old map filed with any number of lot measurements not shown. Because of the activity in that area, the city might have its engineer make an overall study and field survey by which to file an "official map" showing the previously missing information.

There is a special type of map used in many areas under different names, such as, "Licensed Surveyor's Map" or "Land Surveyor's Map" or "Record of Survey Map" or "Certified Survey" which, by whatever name adopted, is in effect a public record of a private survey. This private survey, as such, generally does not change or create lines of title. There are, however, four exceptions to this:

1. If this type of map is requested by the court to be prepared and filed in a Decree of Partition and the court tells the surveyor how to divide the land. This is a form of subdivision without governmental procedure by the executive branch.
2. If a part of an agreement line surveyed and shown on the map is then cross-referenced into and with whatever official record is recorded on the agreement line between adjoining owners.
3. When it is used as a correction map showing errors and/or omissions on a previously filed map. The legal acceptance of this process should be verified before using it.

4. Sometimes, if permitted by law, when it is used in lieu of a subdivision map.

Whether you can use the "public record of a private survey" type of map as a vehicle for the transfer of title by referring to a parcel shown thereon depends on the code or other statutes in your state. A close check on this matter should be made before such a map reference is used in a description, or, if found in an already recorded document, before it is accepted as bona fide.

Normally, it is not desirable to use assessor maps for reference in the conveyance of title to property. However, there are some states which, by law and with the execution of a certain procedure, permit the assessor to file his maps in the recorder's office. In other states, the auditor may request the county surveyor to prepare a plat showing physical locations of boundaries (in contrast with the deed) and then by approval of the city council, the plat is duly recorded. (Some areas require the owners to sign the plat while others do not.) This, in turn, permits reference to a parcel shown on such a map or plat and, with the proper recording reference, it is acceptable for the conveyance of title to that parcel.

There are 30 states in the domain of public lands that have had their survey work executed under the 1785 Ordinance which established the United States Rectangular Survey System of Townships and Sections. The official plat of these surveys is known as the township plat and it is filed in the district land office. Besides the subdivision of the township into sections and lots, these official plats are also used to show mineral claims and tracts of lands that were applied for prior, and, in some cases subsequent, to the subdivision of the township. The patent that is issued for the sale of any of these lands refers to the "official plat" that shows the land described therein and although it is also called a "township plat", the latter wording is not used in legal descriptions.

Patents that were issued by the United States Government in conformity with the grants of the Spanish and Mexican Governments to individuals according to their diseño and expediente included the map of the survey, and when recorded, became public notice of the confirmation of the grant.

In a few areas, provision is made by law for the filing of a correction certificate, an affidavit or an amending map to correct errors on previously recorded subdivision maps. This does not include any redesign of the lot configuration; it is for the sole purpose of correcting drafting or typographical errors. In some cases, this is applicable to other types of maps also.

Sometimes the court will require a map, or maps, to be made by a registered professional surveyor or engineer for a specific case, and it may or may not carry authorization to be recorded in a public office. This is different from the type referred to above under the licensed surveyor's map procedure.

Remember that although a map may be a public record, it is not by that classification alone necessarily constructive notice. Read your code or statute carefully and make definite note of which maps fit the category of constructive notice because they are the ones acceptable for reference in a legal description of land.

Of course, there may be instances where the document containing the description referred to as "Higgin's south forty acres" resides in an old shoe box on the top shelf of Grandpa Higgin's clothes closet because he does not know the value of recording his deed.

Via Extrinsic Sources

In a different vein, you may find descriptions with which you must deal, having not only no caption but no ties to record information and/or no sense in the manner of wording used to supposedly describe the subject property. The following was in a deed recorded in 1947.

> "Pt of Lt 3 Blk 14 Electric Railway Hm std Ass'n 14/27 MR 1011 ½ S. Harvard Blvd. Being 25 ft on a ln running N8° S on cement sidewalk 140 ft N from incinerator to ln of W side lt th 145 ft E on the wooden fence to line of starting."

Here you do have reference to a recorded tract map, but what is meant by "N 8° S"? Which direction is the "25 ft" which presumably starts "140 ft N from incinerator"? Which way does the west line of the property in question go? If there is more than one "wooden fence," which one did they mean?

If you can find surrounding surveys, there might be enough information to help you pinpoint the poorly referenced lines but absent surveys, what would you do? With or without field support, a search should be made for adjoining records by which all the parts of that Lot 3 could be put together. If that is not sufficient, then perhaps it would be necessary to get the description of every deed issued in the entire Block 14 and lay them out exactly as each description reads using ties where cited. In the absence of ties, do not assume a coincidence of boundary lines; if the drafting produces an overlap or a gap, show it the way it is. With all of the facts surrounding your problem parcel in front of you, the relationship of it should be more evident and from that you may be able to create a suitable description.

An example of a road deed of earlier vintage reads:

> "A strip of land sixty feet in width commencing at the north end of where the new bridge across the Santa Ana River at Yorba is to be built, location having been made by the Board of Supervisors, and the contract having been let for same. Said bridge being located at the center of the south end of said strip; said strip extending sixty feet in width in a northeasterly direction until it intersects the old road at the corner of the cypress hedge at the present home of the grantor."

This has no record information whatsoever; it would be necessary to get the grantor's deed, any other deeds available and all the survey notes and maps which were made for the road and the bridge. Availing yourself of all of these facts, you hopefully can relate the position of the road and bridge to ownership lines.

Thus it is that extrinsic help is sometimes needed in order to decipher the problem.

DEED MAPS

Another approach to support information is by the creation of it ahead of time instead of having to look for it after the fact. One form of this kind of help is found in the use of a map attached to the document.

Many legal descriptions are lacerated not only by the writer but also by the typewriter, and having never been proof-read, any kind of a problem can be created which like a dead end street shows you nowhere to go; consequently, there is no recourse within itself by which to find a correction. This is the juncture at which a map or plat attached to the description showing the lines supposedly described can be of tremendous assistance.

For a simple illustration, it is so easy to find the correction of a mistaken direction when you can *see* on the map that the bearing cited in the description as "S 53°37′45″ W" is a line shown in a south*easterly* direction, so, obviously the bearing should be, "S 53°37′45″ *E*".

It is indeed unfortunate that there are so many documents on record containing long metes and bounds descriptions without the benefit of a map attached and made a part of the instrument, because in such cases, it requires the expenditure of considerable time and effort (which represents money) to draw the map, and this often requires the creation of a mathematical traverse to verify the accuracy of closure in the case of a closed area or the establishment of coordinates in the case of a strip description.

If a person who is experienced in field surveys, writing descriptions, and drawing maps uses maps for comparison and verification of the other facets of surveying (and I can assure you they do), then how much more valuable would a map attached to a legal description be in helping anyone less experienced in such matters.

In Corpus Juris under "Boundaries," it states, "Where in a description of land a reference is made in the instrument by which it is granted to a map, plat or field notes, together with calls for courses and distances, the latter will generally yield to the former, in case there is a conflict or discrepancy between them, and will control the location of the boundary." A good number of court cases from across the country are cited in support of this statement.

A common form of statement used regarding the inclusion of a map with the document is, "all as shown on the map attached herewith and made a part hereof." This need not be limited only to deeds, it can be just as applicable to mortgages, leases, wills, etc. The more complex the description, whether it be covering long, twisting easements or multi-sided areas, the more desireable it is to have a map or plat of the area including ties and details attached to and made a part of the description. Even less detailed descriptions can sometimes be helped by a "map attached."

ASSIGNMENT

Go to the various offices where the different types of maps are deposited. Look at these maps and make yourself acquainted with the labels on the books in which they reside, or any other form in which they are filed, such as numbered tubes, the type of index that is used for finding them, and the kind of reference to be used when you incorporate them in your description. Ask for the proper wording to use in reference to each map because some may be "filed" while some may be "recorded" according to the form designated in the code, and this is a part of making the correct delineation.

For quick retrieval from your notebook, make a list of these maps and give the correct reference for each one. Use double posting; that is, one by book and page numbers and the other by tract name or number.

PROJECT

Choose at random a parcel of land in your county, the owner of which you do not know, and go through the process of looking at maps, going to the assessor's office and going to the office of the public records to find the name of the owner and the description of the land. Having acquired this information, verify the facts by checking through other avenues of information. If you find a difficult description go through the procedure of extrinsic research as outlined above.

PROBLEMS

1. Name the two public offices in which you can find copies of legal documents and owners' names and addresses.
2. Name six offices where you can expect to find maps which will assist you in research of title and survey information.
3. Name four instances where a "Licensed Surveyor's Map" or "Land Surveyor's Map" or "Record of Survey Map" or "Certified Survey" does create, establish or change lines of title.
4. When a description is absent vital information, what do you try to do to fill the void and determine the proper outline?

Chapter 3
Description Fundamentals

Introduction

A description of the outline of a certain area is the proper grouping of words which delineates one specific piece of land and which cannot apply to any other piece of land. Although the definition is correctly stated as "land" description, it is generally called "legal" because it must stand up under the law and litigation.

There are generally five types of descriptions:

A. A series of lines around the perimeter of an area known as a metes and bounds description; "metes" means bearings and distances and "bounds" means and refers to monuments both physical and legal.

B. A boundary description stated totally by reference to other parcels already on record in the public repository. It merely ties by reference to the deeds of the owners adjoining each boundary line of the property being described.

C. All or a part of a section in Public Land Area.

D. A lot in a subdivision as shown on a map.

E. Strip descriptions for right-of-ways of pipelines, roads, etc., which may cross many parcels.

The main parts of a description are the caption and the body, and sometimes there are qualifications such as exceptions and/or reservations. These parts are like a camera; you point the camera in the general direction and it embraces the limited area within which you are dealing, in other words the caption, and then you bring the camera into focus on the subject itself and this is the body of the description that details every part.

A caption possesses the inherent function of limiting the title within which it, and/or the detailed description following it, may operate. Thus, "That portion of Lot 2 . . . . . .described as follows:" or "That portion of the southeast quarter of the southwest quarter of Section 13 . . . . . .described as follows:" prevents any part of the detailed metes and bounds which may be outside of the confines of the caption from carrying any transfer of title with it. (Piper v True 36 Cal 606.)

See Figure 1 wherein Lot 2 is the rectangular shape and the metes and bounds follow the irregular shape with arrows. By the caption of "That portion of Lot 2 . . . . . .described as follows:," the hachured area included within the description but outside of Lot 2 would not pass in title. Neither of course would title pass to the triangular areas within Lot 2 but not included within the described area.

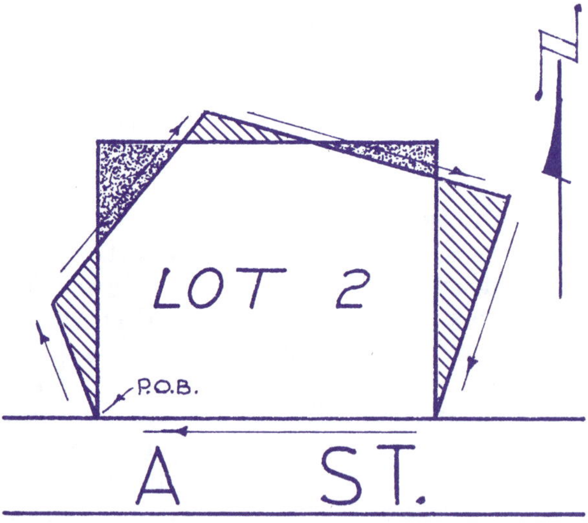

Fig. 1

The preferred method of writing descriptions is using the best type or combination of types and parts that will give the clearest and shortest description possible. In some cases, the entire description can be included within the caption. Your slogan should be "concise clarity without ambiguity." However, it is advisable to be certain of sufficient detail and words so you do not have any possibility of misunderstanding. It is far better to add a few words than to be short of the meaning.

A few examples will help to illustrate some of the problems which arise from insufficient explanation. One from a northeastern state told of:

> "Beginning at a hole in the pond; thence easterly three hatchet lengths; then southerly six hatchet lengths . . . . . etc."

The "hole" in the pond was the spot which was too deep to freeze over in the winter. The "hatchet length" was the distance a woodsman's ax could normally be thrown.

Another deed covering a transaction of land in a part of the old Mexican territory read:

> "Beginning at the northwest corner of Pablo's corral; thence northwesterly three smokes to an old willow tree; thence southwesterly five smokes to the Zanza. . . . ., etc."

How far is one smoke? Exhaustive inquiries of old-timers unfolded the story of a caballero on horseback who smoked cigarettos at a leisurely pace enroute and he consumed three of them during his ride from the corral to the tree. . .and so forth and so on.

Both of these descriptions were satisfactory at the time and in the place they were created, because the people involved knew what the wording meant. It is sometimes very difficult for one, a century later, to know what they meant, and by our standards of accuracy today, these designations for distance are very inept.

Another deed of record in the old sleepy pueblo of Los Angeles read:

> "Beginning at the north end of an old adobe wall now removed. . ."

Here the people on each side knew where the wall had been so they knew where the described line was on the ground, but today there is no physical evidence to support the position of that line.

The following appeared in a deed recorded in Hartford, Connecticut in 1812:

> "147 acres, 3 roods and 19 rods after deducting whatever swamp, water, rock and road areas there may be included therein and all

other lands of little or no value, the same being part of said deceased's 1280-acre colony grant, and the portion hereby set off being known as near to and on the other side of Black Oak Ridge bounded and described more in particular as follows; to wit: Commencing at a heap of stone about a stone's throw from a certain small clump of alders, near a brook running down off a rather high part of said ridge, thence by a straight line to a certain marked white birch tree about two or three times as far from a jog in the fence going around a ledge nearby, thence by another straight line in a different direction around said ledge and the Great Swamp so called . . . thence after turning around in another direction and by a sloping straight line to a certain heap of stone which is by pacing just eighteen rods and about ½ rod more from the stump of the big hemlock tree where Philo Blake killed the bear, thence to the corner begun at by two straight lines of about equal length which are to be run in by some skilled and competent surveyor so as to include the area and acreage as hereinbefore set forth."

A description from a will in the County of Prince William in Virginia read thus:

"Beginning at a pine; thence south 23½ west 87.2 poles to a bunch of dogwood sprouts by three red oak sprouts; thence south 45½ west 89 poles to a small hickory and several marked pointers . . . "

These have references to various points or things obviously difficult to recover. This indefiniteness is not confined, however, to the east nor the area of the thirteen colonies; witness the following descriptions as parts of ranchos out west:

"Beginning at a post in an irregular mound of earth and bones of slaughtered cattle at a place called Corral of the Mission about four chains north of the point of a Cienega . . . "

and:

"Beginning at a post in a mound of sand at the high water mark of the Pacific Ocean nearly opposite the center of an island one mile long . . ."

Where are they, and how do you orient these points?

You too no doubt have a collection of memories of other difficult descriptions.

The lesson to learn here is that correct definitiveness is of extreme importance.

WORDS AND PHRASES:

Besides using a certain quantity of words, one must apply the proper qualifications in order to obtain the necessary results. For one writing a description to use words that connote a picture of the situation from his viewpoint but which, even though they may be concomitant, do not clearly and legally denote the facts introduces various degrees of ambiguity. In the terms of law, *latent ambiguity* is an uncertainty which does not appear upon the face of an instrument but arises from evidence aliunde of the words themselves. This is in contrast with *patent ambiguity* which is obvious from the words expressed.

A thorough understanding of words and combinations of words from the legal standpoint is not only a great assistance but is a must in preparation for the writing of descriptions. In some cases, it is just a matter of using good English grammar.

For legal determination and interpretation of words, four dictionaries are recognized as authoritative; they are Ballentine's, Black's, Bouvier's and Burrill's. The American and English Encyclopedia of Law is also used especially for older descriptions. You should have access to at least one of these for making any decisions on the correct meaning of any questionable words.

Adjacent; Adjoining; Contiguous; and Contiguity

Consider these words which are often used interchangeably and yet they are not strictly synonymous.

According to Black's Law Dictionary, the word adjacent means "lying near or close to; . . . implies that the two objects are not widely separated although they may not actually touch."

You have seen in different cases this word "adjacent" used with the intent of meaning that one parcel of land is connected with another parcel such as an easement with a parcel of land. But since the word in its true sense means "lying near or close to," you thereby have a condition where the easement is not necessarily attached to the piece of land to which it is intended to serve.

Again referring to Black's Law Dictionary, the word "adjoining" is "in its etymological sense, touching or contiguous as distinguished from lying near to or adjacent."

The word "contiguous" is described in Black's Law Dictionary as "in close proximity; near, though not in contact; neighboring; adjoining; in actual close contact; touching; bounded or traversed by."

The word "contiguity" is described in Bouvier's Dictionary as one parcel contacting another.[4] It appears that the noun "contiguity" carries a better connotation than the adjective "contiguous" insofar as its descriptive application to one parcel serving another is concerned.

It would further appear that contiguous is not the most desirable word to use because of the contradictory definitions; therefore, if the easement is intended to touch the subject property, it would be better to use the form:

"a 6-foot easement adjoining the southerly line of said land."

However, if the following form were used, you would have no argument whatsoever that the easement is joined at all points along the line:

"together with a 12-foot easement, the northerly line of which is *coincident with* the southerly line of said land."

The phrase, "coincident with," assures the conveyance of the meaning of constant and continuous contact along the line.

Conveyed and Described

These two words are sometimes interchanged or used with the idea of carrying the same meaning when, in fact, they have different meanings; in other words, "conveyed" does not convey the same meaning as "described."

When you read a phrase such as, "thence southerly to the south line of the land described in deed to Jones;" you must look at the description in his deed and determine that south line *as described* and use it. If, on the other hand, the phrase reads, "thence southerly to the south line of the land conveyed to Jones;" you must do more than read his deed. It may even be necessary to go to an attorney or a title company to get an opinion as to the limit of "the land conveyed."

Referring back to Figure 1, if Jones' deed said, "Lot 2 in Tract Number _____, etc.," then "the south line of the land described. . . ." is the south line of Lot 2 as well as being the north line of the STREET.

Contrariwise, if the STREET is a normal interior street in that tract, then, on the basis of lots carrying an underlying interest in the adjoining street, "the south line of the land conveyed. . . ." would be a line *in* the street. This normally would be the center line but it could be a line in some other position within the street, depending on the

[4] "Contiguity" means that the boundaries are "coterminous" according to the following court cases: Rafferty vs. Covina, 133 CA 2d, 745, 285 P 2d, 94, and Irrigated Valleys Land Company vs. Altman, 57 CA 413, 424, 207 P 401.

circumstances that created it and this is why you need an attorney or a title company to research the background of the street area. The same reasoning applies to a field survey of a parcel referring to "the land conveyed" in order to determine what to show on the map.

Furthermore, this does not mean that the description on the deed, "Lot 2 in Tract Number _____, etc." does not carry its underlying interest in the adjoining street; we are discussing here as to the *reference* of "land described" vs. "land conveyed."

More on the analysis of other words used to convey, or not convey, interest in the street will be discussed in Chapter 9. The proper use of "conveyed" or "described" in the creation of descriptions will be treated in Chapter 11.

Along

"Along" is sometimes used in a collective sense to cover an aggregate total distance of several segments. In this use it generally carries a correlative statement such as "thence along the easterly lines of said land, northerly 654.6 feet to the northeasterly corner thereof." In this sense, it is not a desirable form for two reasons. First, you have no reference as to the segments and if it is a boundary, you must know the facts; second, if any of the segments depend upon ties or monuments, physical or legal, then you do not have the control necessary to establish that line.

When "along the bank or shore" is used in reference to a specific line in a description in a deed, it may mean "by, on or over," according to the subject matter and context according to Corpus Juris 9. In Howard v. Ingersoll, 13 How. (U.S.) 381, 417, 14 L. ed. 189, it was held that one of the boundary lines defined as extending "along the bank" of a river and running "on the bank" limited the conveyance to a line drawn on the bank and did not extend *ad flum aquae* to the center of the stream. Similarly, in another case, describing the boundary of a mill lot; thence southerly "along the shore" of said river, it was held that the word "along" as there used meant a boundary line drawn along the shore at low water mark and the grantees took no title to any part of the bed of the stream.

Contrariwise, in Walton v. Tifft, 14 Barb. (N.Y.), 216, it stated that where a grant of land is described as touching the water of the river and running thence along the river as it winds and turns (and being non-tidal) and the parties did not expressly except the river, one-half of the bed of the stream is included by construction of law.

By a line, or upon a line, relates to being on or along a line. When Along, By or Upon refers to a strip of land, a right of way or road, without qualification, it carries to the thread or center of the strip.

Between, By, From, On, To

These words have been cited in many cases as being determinative of a boundary by exclusion. When a line in a description goes "between" or "by" or "from" or "on" or "to" a boundary, then there is established a limit to the subject property. Under a different situation, however, in Leary v. Mayor etc of Jersey City, 189 Fed. 419, 428, the court stated, "The words 'to,' 'on,' 'by,' 'at,' 'along,' a nontide water stream presumptively carry title as far into the stream as the grantor possesses."

Direction of Segments vs. A Whole Line

Sometimes you may have reference to an "easterly line" considered collectively although it has many segments of different bearings, including some which may be nearly at right angles to the north-south direction of the easterly line. The courts have held that a group of consecutive lines with different bearings, including some which oppose a generally north-south line of property, will be all inclusive within that statement, in spite of the fact that there may be some contrary directions. (See Figure 2.)

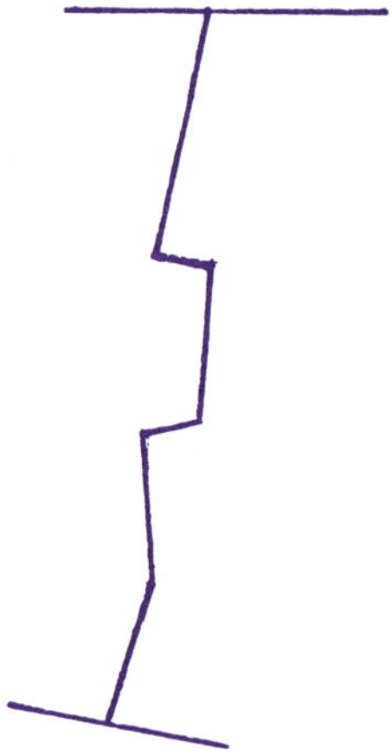

Fig. 2

Due vs. True

Due east, due north, etc., have been used many times but what does it mean in contrast with, or in comparison with, true east or true north, etc.? In the Manual of Instructions for the Survey of Public Lands by the Department of Interior, Bureau of Land Management, "true" is referred to as the direction of cardinal lines and true north is specified as astronomical north. It further specifies that the direction of all lines will be "stated in the direction of angular measure referred to the true north or south . . . " It does not use the word "due." (Section 40, page 40 of the 1947 Manual and Section 2-17, page 19 of the 1973 Manual)

In the new manual of 1973 on page 5 it says under the rules of survey that "the north-south lines shall be established according to the true meridian . . . . " And on page 61 it states, "A Principal Meridian is intended to conform to the true meridian." As far back as Tiffin's Instructions for Deputy Surveyors in 1815, they were admonished that "the north and south lines be run according to the true meridian" and "to run a true north course." So the basis for the use of the word "true" has considerable historic background.

The word "due" is not titled in Black's Dictionary; however, following, "NORTH," he states, "Means due north. Same with *northerly*." This indicates that "due north" like "*northerly*" is a general, rather than a finite, direction.

"True" as applied to a direction in a description is considered to be an astronomical direction. The courts have held that such a line is to be established by derivation from solar or Polaris observations. In addition, if sufficient geodetic control monuments exist in the vicinity of the property in question, orientation to the true north may be accomplished by tying to them.

To show how careful you must be with the interpretation of words such as "due" and "true," consider the case of Richfield Oil Corporation vs. Walter M. Crawford, et al.[5] Extrinsic evidence by surveyors, engineers and others was admitted and from all the facts considered, the court decreed that the word "due" changed the base of bearing from "North relative to a given line" to "astronomical North" as determined by a star observation.

In conclusion, it should be stated that although there will be many times when you must analyze the effect of the use of the words "due" and "true," you would do well to limit yourself to the one word "true" in creating descriptions, and preferably only in those special circumstances where it is demanded by survey.

Either

This word is too often used in describing a strip of land as being "20 feet on either side of the following described line:." The etymology of the word indicates a choice of one thing or another so in the above form it would mean that you have 20 feet on one side *or* the other side of the described line. From this standpoint, there are two problems.

[5]249 P 2nd 600. A brief of this matter appears on pages 157, et seq. in Vol. XVII No. 2 Quarterly Journal of the American Congress on Surveying and Mapping by Curtis M. Brown.

First, on which side does the 20 feet exist, and second, how do you account for the other 20 feet which is apparently intended?

In the place of "either" use "each": "20 feet on each side of the following described line:" will cover a 40 foot strip with a description of the center line given.

Excepting and Reserving

In the old days "excepting" and "reserving" were used together or interchangeably. This is confusing because of the difference in the meaning of each word.

When you read "excepting and reserving a roadway over the south 30 feet," does that mean the "south 30 feet" is cut off from the fee conveyed or is it merely a reserved right for a roadway?

According to Webster, the word "except" means to pick out or leave out; or exclude; to omit. It is a cutting off of a portion of that which has been described.

"Reserving," in contrast, does not mean cutting off but it means taking and keeping a certain right from a part of the whole area which has been given in the deed. Or, as the courts have said in some cases, you give the fee away and then you take back a certain right. Again referring to Webster, "the act or fact of a grantor's reserving some new thing out of the thing granted." By summation, the fee title passes and the reservation retains some right or rights to be exercised over this particular part.

When the reservation is *first* created, it appears at the end of the description or after it in a separate paragraph. If the reservation is already described in a document of record, then all subsequent references to it will be treated in a different form of wording. It will not appear after the description as a reservation; it will be stated in terms, "subject to the reservation of. . . . " and the facts will be cited as in the document where it was created. It is desirable to also include the date of recording, the document in which it was created and the book and page reference. For example, if an easement is reserved over the east 10 feet of said land for driveway purposes, then a subsequent description of that area should show the easterly 10 feet as subject to the reservation created in the former deed giving the date and reference for it.

There is a rare instance, however, when you may properly use the combination of these two words but it is in different context. If you buy a piece of land which carries "EXCEPT therefrom 50% of all oil, hydrocarbons, minerals and gas." and you then sell it, reserving to yourself the other 50% of all oil, hydrocarbons, minerals and gas, a subsequent statement could be "EXCEPTING and RESERVING therefrom all oil, hydrocarbons, minerals and gas." This would be a correct statement of fact because of the two previous and different documents.

Except the interest in . . . .

In the acquisition of strips of land by political entities, sometimes the document clearly states it is a RIGHT OF WAY DEED and at other times it is obviously a Grant Deed for fee title. There are instances, however, when a Grant Deed form is used and *supplemental* phrases are added, such as "for road purposes."

It is not the purpose here to dissertate on the law regarding the interpretation of such wording as to whether they pass fee or easement. The objective is to alert you to the possible question, if it is not clear-cut, and the way to write around it for the protection of your client and yourself.

If any document conveys a strip, or any shape, of land to a political entity for public use, and the wording of the grant is not clear as to fee or easement, and that area is a part of your property, you can follow your description with,

> "EXCEPT the interest in that portion thereof described as follows:"

By the use of this form of statement, you are saying that there is an interest which you do not wish to classify in that particular area. In this way the decision is left to the court.

Front and Rear

"Front" or "rear" are not satisfactory words to use because of their indefiniteness. The outstanding example of this problem is the application of these two words to a corner lot where you would not know which line would be front or rear. The use of compass direction is preferable and leaves no question. "The north and south line of said lot" or "the southwesterly and northeasterly lines of said lot," give a clear and understandable position of the lines.

In a straight line

The statement "in a straight line" is usually unnecessary because the shortest distance between two points is a straight line. An exception might be if you were following a fence line which was not straight and you wanted to be certain that the described line followed on a straight course between angle points rather than the wandering fence.

More or Less

This statement is ofttimes used extraneously. In one description it followed every course and when the matter was taken to court, the judge knocked out all of the statements of "more or less." The theory expressed was that inasmuch as they were used throughout, they carried no meaning whatsoever.

When a course ties to a matter which is on record, or ties to a monument, the tie will hold regardless and so it is unnecessary to say "more or less." Also, when this statement "more or less" is put in the last course when going to the point of beginning, it has no value because the call is mandatory to go to the point of beginning regardless of the bearing and/or distance. If a course recites, "thence to the northwest corner of Forbes' land," the surveyor must first establish that northwest corner and then measure to it and report whatever the bearing and distance is. Here again, there is no reason to use "more or less."

A good place to use this statement is where you know, or strongly suspect from extrinsic evidence, that there is a problem. For example, if you have no survey information relative to a certain line but there are several conflicting distances available from foreign sources, pick the one that fits the most logical deduction of the facts and follow it with "more or less" to indicate the existence of a problem.

What does the citation of "Westerly 567.463 feet more or less" mean? Is it really that accurate? If the measurement was not done that accurately in fact, why was the distance given to a thousandths of a foot? If you had information that this measurement was in reality only within approximately 10 feet of the correct answer, then the above distance should have been stated 570 feet, with no decimals, and the use of "more or less" would be apropos to indicate the knowledge of a problem in that area. This signals a warning to anyone reading the description.

Parallel

In the use of the word "parallel" the correct preposition to be used with it is "with"; you are never parallel *to* something, you are "parallel with" a line or a wall or a street, etc., etc.

Parallelism is not limited to the confines of the lines under discussion. For example, see Figure 3. In such a case, it is not necessary to say "parallel with the prolongation of such and such a line." The courts have held the fact to be sufficient that the prolongation holds an equal distance condition between the prolongation of the two lines and is sufficient without having to say "parallel with the prolongation" of the other lines. In the illustration, the description, "the north line of Parcel A being parallel with the south line of 'U' street" is sufficient even though the two lines are not opposite each other.

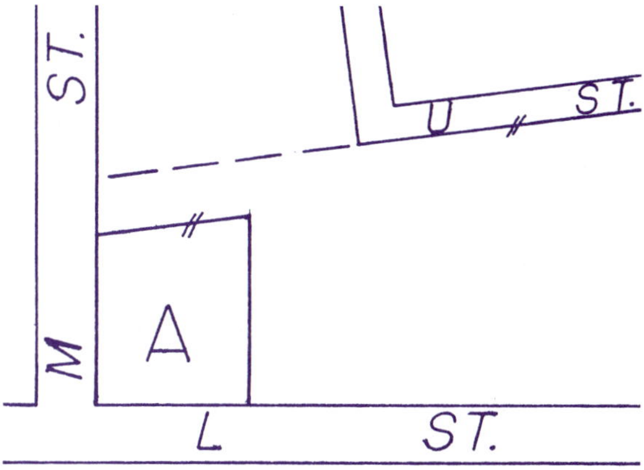

Fig. 3

Point

"To a point," "at a point," "to the point of intersection." The end of a line is a point so it is not necessary to state the fact. An intersection of two lines is a point. It is really unnecessary to make a point of "to a point" or "point of intersection" *unless* there is a lateral tie that must be held in which case the "point" becomes a qualified control.

Portion

This is used when referring to a part of the whole area. Sometimes it is a part of a single parcel and other times you have parts of several parcels or several parts of one parcel. The point to remember is that, as in English grammar, the wording should conform with either the singular or plural situation. In other words, when you have only one part you say, "that portion of . . . " For example:

"That portion of Lot 5 . . . . described as follows:"

or:

"That portion of the southwest quarter of Section 12, Township 4 South, Range 2 East, of the Navajo Meridian, State of Arizona, described as follows:"

or:

"Those portions of Lots 7, 8 and 9 . . . . described as follows:"

or:

"That portion of Lot 183 of the Newport Beach Tract and that portion of Section 25, Township 6 South, Range 10 West, of the San Bernardino Meridian in the County of Orange, State of California, described as a whole as follows:"

Position (Words, figures and phrases)

The juxtaposition of words and figures in a description is very important. Very often the closeness of figures and words to each other creates a question as to the meaning. If your description reads "west along the north line of Lot 3 201.74 feet," does it mean west along the north line of Lot 3, 201.74 feet, or does it mean west along the north line of Lot 3, a distance of 201.74 feet? It is far safer to say, "West 201.74 feet along the north line of Lot 3." This same principle applies to curves as well. In every case, if you will put the distance and direction together, you will have a more satisfactory understanding and it will be safer of interpretation. For example:

> "North 322.63 feet along the west line of Lot 3 to the beginning of the curved southerly line of A Street concave to the south and having a radius of 1,000.00 feet; thence easterly 164.89 feet along said curve . . . . etc."

or:

> "Thence along the southerly line of the railroad right-of-way, south 89° 15′ 46″ east 600.00 feet; thence leaving said right-of-way on a deflection angle to the southeast of 36° 10′ 12″, south 53° 05′ 34″ east 289.54 feet . . . . etc."

Prolongation

Projection, extension or continuation are sometimes used in lieu of prolongation. Continuation is used with the description of curves and means extending, or the extension of, the arc of the curve. Prolongation, as applied to the curve, means the extension of the *tangent* line of that curve at that point. In other words, a straight line is "prolonged" and a curved line is "continued."

Prolongation, projection and extension may be interchanged as applied to a straight line.

Right or Left

Although the use of these words has been common over the years, they are not desirable from the standpoint of safety or even clarification. The use of compass direction is much better and more reliable. The use of these words will be enlarged upon and illustrated in Chapter 4.

Running

Although this word has been used for years, it adds nothing to the meaning of the description. It would be well to classify it as verbose and omit its inclusion.

Said and Aforesaid

Oftentimes, unnecessary repetition is prevented by the use of the word "said." It refers to an immediately preceding matter of the same subject; never to a subsequent matter. For example:

> "thence north 48° 26′ east 462.25 feet to the west line of the land described in deed to Tom Brown recorded in Book 928 at Page 32 of Deeds; thence southerly along said west line to the southwest corner of said land."

In this instance, the word "said" was used twice in a manner that saved repeating the reference to Tom Brown's deed and its recording reference. Its usefulness for simplicity and certainty is unquestioned.

"Aforesaid" is nearly synonymous with "said," although it may cover a little broader scope. Its use is not as common nor recommended.

Subject To

Whereas the statement "granted to" or "reserved from" or "the reservation of" refers to the instant condition which is being created, the words "subject to" refer to an already existing, a previously created, situation. Where a document is already on record and the grants or reservations are being carried forward, the subsequent reference to that must be made in the form of "subject to." For example, assume that you sell the north 50 feet of your lot to Murray and grant him an easement for ingress and egress over the east 12 feet of the remainder of your lot. When you subsequently sell that remainder of your lot; you show it "subject to an easement for ingress and egress over the east 12 feet thereof."

Thereof

Refers to a previous item in the description. In case of an exception to a certain part, i.e.: "Lot 4 in A's Tract . . . . EXCEPT the south 25 feet thereof."

Sometimes used incorrectly, grammatically speaking, i.e.: " . . . . along east line of said Lot to a point distant 78 feet northerly from the southeast corner thereof." *Thereof* refers properly to the *line,* not the lot; there is no corner to a line.

Relationship Between Descriptions and Surveys

In the analysis of descriptions, and especially boundary problems, a decision must be reached as to what really controls. Land descriptions and land surveying are interrelatedly pertinent in the determination of boundaries of areas. The positioning of any boundary is dependent upon whether there is a description of it already on record in the public depository of records, or whether there must be survey work performed to establish the lines and subsequently draw a description for them. In either case, it must be recognized that inasmuch as the physical monuments placed for marking corners at the extremities of the lines, and occasionally at intermediate points, are subject to movement, accidental, intentional or by natural processes of the earth, the created description must incorporate either ties to the most satisfactory physical monuments available, or sufficient and applicable ties to matters of public record, or both.

When the record is properly written or mapped, it becomes the controlling factor in any location or restoration of a boundary which may not have been monumented, or where the monumentation has been physically destroyed. "In the absence of monuments or marks, or in the case of repugnancy between them, the determination of what are the true lines called for in the entry, grant or conveyance is a pure question of construction, dependent on the intention of the parties, as shown by the instrument and the circumstances of the transaction."* This is why the preparation of the record should be done meticulously by including references to physical monuments, legal monuments, dimensions, ties to pertinent objects, lines and points.

Positioning of the boundary of a parcel of land is the result of applying a set of record conditions created for that boundary to the circumstances existing in the field. Where there is non-conformity between the two, then the surrounding areas must be combed for any acceptable monuments, lines of possession and accessories after the fact for intrinsic help in tieing them all into a dimensional net. Additional information from the record, quasi public record maps, and even owner affidavits should be brought into a completely corroborative picture. This total picture should then be analyzed to give suitable weight to the various differences and for which the positioning of those lines and corners can accomplish the greatest satisfaction of the most requirements of the combined facts available. This is known as making use of the majority probability of all conditions for the most satisfactory solution possible.

* 9 C.J. Boundaries Section 45.

If, however, there is so much disparity between the facts revealed by the survey and the words in the description that a reasonable answer is inept, then recourse to an agreement line is most plausible. Here, the establishment of the position of a line on the ground by mutual consent of the adjoining owners, the writing of a newly-described line derived therefrom, the preparation of a detailed map, the drawing of an agreement with proper words of conveyance and the official recording of such a document, either including the map attached or with reference to a compatible map on record will accomplish a solution. Very often an agreed line is less expensive, less trouble and far less time-consuming than litigation.

When the description is already on record in a document in the recorder's office, the survey must follow that description in retracing and establishing the lines therein on the ground. When there are ties called for in the description to a certain owner's land, or to a highway or to a marked rock or the railroad right of way, then those items control and whatever the measurements to them are on the ground is the answer. Likewise, when the description in the documents refer to facts shown on a map or there is a definite statement of bearings and distances, then they will hold and control the positioning of the lines of survey in the field. If there is a flat call of a distance and/or bearing, it is not prorated.

When a description ties to a monument, that monument controls the survey. When a description ties to the south line of Smith's land, then the survey must establish the south line of Smith's land and go to it. If the description ties to a street intersection, then that must be established first by survey and then the survey of the other lines are tied to it. If the description ties to the west line or to the northwest corner of the southeast quarter of the southwest quarter of the section, then that quarter must be established on the ground first and then the ties made accordingly.

Contrariwise, there are conditions where the survey controls descriptions. When the owner of a colony grant would walk around an area with a buyer and a surveyor and indicate certain trees, rocks, creeks, etc., to be the boundary, the survey of the lines between those monuments would control the description to be written for the sale of that land. Later on when the U. S. Rectangular Survey System was established, the surveys made pursuant to the rules of that System were, and still are, the control for the descriptions used in the conveyance of those lands.

Surveys made for tideland locations and mineral patents control the description that is used for those conveyances. Mining claims which are based on the location of the lode are controlled by the outer limits of the ground location and then that perimeter is tied to section corners for the purpose of describing the area to be patented. By federal law, a survey of public lands must be made and described before the patent can be issued.

Another form is in the development of right-of-ways for pipelines, power lines, highways, railroads, etc., which control the descriptions to be written.

When subdivisions of land are created with blocks and streets, alleys, walkways, and so forth, and a map is created from those surveys, the map of the survey becomes the resulting control of the description for that land on the assumption that the monuments coincide with the facts shown thereon.

Similarly, but in a three-dimensional form, condominiums are created by and controlled by plans and surveys which result in the descriptions used on documents of conveyance.

Aerial easements, such as bridges, are controlled by surveys which are necessary to determine their positions above ground as well as in a horizontal plane.

Similar in form to aerial easements, subterranean easements are controlled by inverse elevations. Usually they are not only required for a subway tunnel but for such appurtenant items as stairways and elevators to and from the tunnel.

In some instances, however, the descriptions for aerial easements or subterranean easements are written first on the basis of construction design determining the needs and then the survey work is performed in accordance therewith.

Avigation easements are a relatively new form of description. The word was coined by taking "navigation" and removing the first letter to make "avigation," which has to do with the flight patterns and the slope easements for take-off and landing into and around airports. These again depend on survey information.

There are other surveys that are used indirectly for control but they seldom appear in a description. Geodetic triangulation points are used for determination of other corners that ultimately result in a local survey control. These lower order control points are, in turn, sometimes used as basis for a legal description. One example of this is the use of coordinate controls for right of way development surveys.

Hydrographic surveys perform a function for vertical control in the task of developing marinas, bulkheads, bridges and even buildings concerned therewith. Again, the information is not a direct control but the development of the project often requires descriptions in three dimensions.

Tri-dimensional descriptions will be discussed in Chapter 14.

PROBLEMS

1. Does a caption limit the boundary of title?
2. Define "legal description."
3. Describe "latent ambiguity."
4. Describe "patent ambiguity."
5. Since "adjacent" means "Lying near or close to," what wording can you use to assure continuous contact between two parcels?
6. Would you accept the statement, "land conveyed" as synonymous with "land described"? Explain.
7. Did the Instructions by Tiffin require the Principal Meridian to be established by "due" north? Explain.
8. Are the words "Excepting" and "Reserving" synonymous? Explain.
9. When the end of a line is a point and the intersection of two lines is a point, is there any need to use the statement, "to a point" at the end of a course and distance?
10. Explain the factors that determine whether a survey controls the description or a description controls the survey.
11. Read the following for understanding and then change words and rewrite in proper form retaining the same facts.

Description:

A 15 foot easement adjacent with Jones' land on the east to benefit Brown in the rear described as follows: (See map on page 3.20)

> Beginning at the southeast corner of land conveyed to Jones, being 330 feet west of southeast corner of the SW ¼ SE ¼ NE ¼ Section 10, Township 4 South, Range 10 East, SBM, County of Riverside, California; thence due north 200 feet; thence east 15 feet parallel to County road; thence south and normal with Brown's land 200 feet; thence west to beginning.
> EXCEPTING AND RESERVING part in County Road.

Brown's property is a cut out of Jones' land described in OR 642-51. Prior to this, Jones owned the land adjoining on the east and when he sold it to Clark, he reserved a non-exclusive easement over "the westerly 15 feet." You are not to use the form of "the westerly 15 feet of Clark's land" in your description because you do not have Clark's description, nor is it good practice to refer *only* to "the land described in deed to ..... recorded in Book ..... Page ....." for the subject property; however, boundary reference is acceptable for a tie to record.

Also, the line between Clark and Jones carries the same description in both deeds so there is no gap nor conflict.

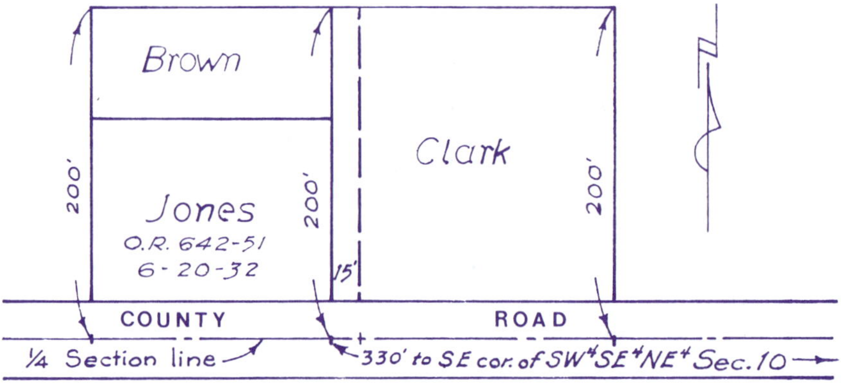

12. Rewrite the following description in better form and fewer words without losing any control.
    In the old days, an owner's land was called his "tract" so in this case, reference could be made to the "Oge and Bond Tract."
    The reason for the two different counties being cited in the description is that, as in many areas, one county has been cut out of another one so that the location of the property is now in the newer county (in this case, Orange), but the map was recorded in the original county (in this case, Los Angeles); therefore, the map reference must remain as per the original location while the physical location is related to its current status.

All that certain piece or parcel of land situated, lying and being in the city of Orange, in the county of Orange and the state of California described as follows to wit:
Beginning at what was known as the northwest corner of Oge's corral being a post on the south side of his road, being more particularly measured as 20 feet from the center of the road and 198 feet east from the west line of Lot 3 in the aforementioned Block "B"; thence from said point of beginning along the exterior boundaries of the land to be described, south along the corral fence 200 feet to the beginning of a curve that is tangent to said last mentioned line, said curve being concave northwesterly with a radius of 40 feet; thence southwesterly along said curve 62.8 feet to a tangent line that is parallel with the aforementioned Oge's road and 240 feet south measured at right angles from the afore described south line of Oge's road; thence west along said last mentioned parallel line 160 feet; thence northwesterly along an old fence 244 feet more or less to an old 4 x 4 marker on the side of said Oge's road; thence east along said road to point of beginning.

— NOTES —

Chapter 4

Determination of What Controls

Evaluation of Items in Descriptions

In the last chapter, the relationship of control between description and survey was discussed showing how they are interrelated and dependent on each other. Now we will consider various items and their value in and to the description.

It has been held by the courts that if the description is surveyable, it is sufficient as between the parties. A boundary is considered definite and certain when by survey it can be made certain from the deed.

If a description is insufficient and ambiguous or has double meaning, it can cause delays, disputes and possibly costly lawsuits and governmental hearings. Any change by agreement, judgement or understanding between the parties involved as to different locations of the lines on the ground as compared with, or in contrast with, the lines described in the deed, lease, mortgage, etc., requires a *new* document in the office of the public records in order to give proper notice to the world of the new lines in the eyes of the law. In land description practice, this procedure of creating a new document for the new lines and making a public record of it is known as perfecting the title.

While reading descriptions you may have wondered sometimes which items really had authority to control over other conditions. Where a call to a boundary deed also mentioned a post that did not agree by distance, which one should hold?

In order to have some semblance of relationships in the value of various parts of control in descriptions, there has been established an order of superiority of calls (sometimes called elements), commonly known as "rules of construction." Although the various codes state them in different ways, the following is a general summation of the ideas expressed:

1. Monuments, which are natural, such as rivers, lakes, oceans, etc., or artificial, such as stakes, iron pipes, cement posts, etc., or legal, referring to matters of record or some combination of these kinds of monuments which are recited. The recital of these in descriptions adds to the supremacy of them over other conditions.
2. Monuments, physical or record, not recited but disclosed by reference to maps, and/or other documents of record.
3. The recital of a boundary of record, such as reference to a map or a deed.
4. Recited distance and/or bearing ties to record corners or lines not on boundary of instant property.
5. Distances on boundary,* without ties to any of the foregoing.

*In Texas a statute provides that calls for distance must prevail over calls for natural objects in the survey of school lands where they will give the county the benefit of those calls that are most favorable to its interest.

6. Bearings on boundary or angles between lines,[6] without ties to any of the foregoing. (Items 5 and 6 are interchangeable dependent upon circumstances.)
7. Areas stated as a corollary or an addendum to the description. When an area is made specific such as "the north 50 feet" or "the north 10 acres," the status is converted to a dimensional control and rates the same as no. 5 above. In Jackson v. Sprague, 13 Fed. 253 Cas. 7148, the court held that where both the quantity of land and a metes and bounds description is given, the latter will control. Addundum details on these calls will follow in correlative areas.

Some of these items as used in descriptions are locative while others are descriptive. Locative calls are specific as to point locations or line locations, such as "to a 4 x 4 post" or "to the west line of David Brown's land." Descriptive calls carry the more general or directive qualifications and merely point to the areas wherein the specific may be found, such as "to the southwest corner of Alan Smith's land as described in deed recorded in book 738 at page 201 in the public office of records," from which deed the details are obtained.[6]

In all cases where the description is being analyzed as to sufficiency and determination of the particular value, there must be not only a consideration of the superiority of call but, along with this, the principle that those material items that are most certain and least liable to question will predominate.[7]

Although the main consideration is given to monuments because they are assumed to have been placed with the intention of showing the evidence of control, there can be importance given to other calls in relation to the circumstances surrounding them. Monuments as such will not always prevail because of certain conditions that may exist. This will be discussed further in Chapter 8. The courts have held that when there appears to be conflict, the call which defeats the intention and the concern at hand will usually be deleted.[8]

Another approach to the reasoning behind the superiority of

[6]United States vs. Redondo Development Company, 254 Fed. 656, 660 (1918). (New Mexico) and Cates vs. Reynolds, 228 S.W. 695 (1921) (Tenn.).

[7]Sandretto vs. Wahlsten, 124 Minn. 331, 144 N.W. 1089. Pardee v. Johnston, 70 W. Vir. 347, 74 S.E. 721. Roberti vs. Atwater, 43 Conn. 540, Newsom v Pryor 7 Wheat 7 (20 US 1822) 5 L.Ed. 110.

[8]State vs. Sulflow, 128 S.W. 652 (1910) (Tex.).

monuments is that they are indicative of the footsteps of the surveyor and therefore they represent the actual physical placement of the line, whereas the courses and distances used to describe the line are only indicative of the survey.[9]

Further, consideration must be given to the fact that if extrinsic information is necessary to locate the description on the ground, it may be considered unmarketable and uninsurable. In such cases, corrections must be made either through an agreement, a corrective deed, or by court action or both. Parol or extrinsic evidence is not admissible where it adds to or varies the meaning or if it changes any definite call.

Applicability of Codes

A listing of the superiority of calls as a generality was just given but you must remember that the specific delineation and application is contained within the codes of each state and they determine the value and relationship of calls in descriptions. Furthermore, there are rules and regulations in the Federal Government Code which parallel some of the state codes but exert their influence on public lands. Certain codes (both state and federal) hold control over the surveys of land in general and the preparation of maps derived from surveys in particular.

Assignment

Prepare a notebook in two parts, the first of which is to be composed of code sections both state and federal that specifically apply to and regulate surveys and maps and descriptions. In this way you will have a ready reference without having to acquire the entire law library just for those matters that are pertinent to writing descriptions. Make a suitable index so you can retrieve any desired subject matter quickly. The method called "double posting" is very helpful; for example: with the heading of this section, "Applicability of Codes," make one posting of "Applicability" under "A" and another one of "Codes" under "C." In this way, no matter which word comes to mind, you will find it in the index.

Along with the code, there are usually references to court cases and a brief resumé of the application which is helpful in understanding the meaning and the use of it. Laws are made by the legislature and

[9]Millerman vs. Megaritty, 242 S.W. 757 (1922) (Tex.) and Andrews vs. Wheeler, 10 CA. 614 (1909) (Calif.).

the courts interpret their use and adaptability or non-applicability.

As a guide to the creation of this legal reference notebook, some of the specific subjects that you should look for in the state codes are:

Amending Maps	Land Surveyors Act
Certificates	Licensed Surveyor's Maps
Condominiums	Maps (general and specific)
Coordinates	Monuments
Corner records	Official Maps
Dedications	Parcel Maps
Exclusion Maps	Record of Survey Maps
Final Maps	Reversion to Acreage Maps
Revocations	Rules of Construction
Subdivision Map Act	Tentative Maps

This list will not be the same in every state, but it covers the general subject matter with which you should be concerned in writing descriptions.

The second part of your code section is to be concerned with the federal code in which you should look for the following subject matter:

Boundaries	Public Lands
Condominiums	Railroad right of ways
Highways	Rules of Construction
Maps	(Legal Descriptions)

Of course, any other topics that you find which are pertinent to your activities should be added to this group.

Copy all for which you may have any need and provide for addendums because changes are made by the legislatures in every session and you should keep up to date with them.

A good memory of the knowledge contained in these codes as well as a ready reference which can be pulled out and set in front of you is a most valuable asset. When a client hears you recite a code reference, or is confronted with a copy of the code relative to the problem at hand, he learns to respect your judgment and decisions.

Each part, state and federal, should carry a supplement of court cases applicable to the subject matter therein.

Basis of Bearings

A "basis of bearings" is the determination for the orientation of angular relationships of lines in a description or on a map. The map usually has a note which states the particular line upon which the bearings have been based. This is usually referenced to a previous map, although, in exceptional instances, the reference may be to a line cited in a deed. Occasionally a bearing will be derived from geodetic control points, or, in some instances, from a solar or Polaris observation. In a description, the basis of bearings, sometimes stated as base of bearings, should be established by the recitation of a control line in the first course. An example of this would be:

"Thence North 0° 15′ West along the west line of said section,"

or

"Thence along the North line of Lot 9 North 89° 26′ West . . . ."

This may be sufficient for an entire description or you may find or use subsequent changes in the base of bearings by references to other lines. For example, when you tie to a line of an adjoiner which in turn is based on a different bearing; then you recite that fact and from there on, your description is based upon a new (that is, new in that particular description) bearing taken from the one given in the adjoiner deed. For example:

> Beginning at the southeast corner of Lot 12 of Livingston's Tract as shown on a map recorded in Book 36, Page 24, records of said county; thence North 0° 48′ West 128.00 feet along the east line of said lot; thence North 38° 26′ East 116.74 feet to the south line of the land described in deed to William Ferris recorded in Book 675, Page 58 of Deeds; thence on the bearing in said deed, South 89° 36′ East 198.32 feet along said south line to a 1″ iron rod; thence South 0° 24′ West 219 feet more or less to the easterly prolongation of the southerly line of said Livingston's Tract; thence westerly along said prolongation to the point of beginning.

Although this is a valid description from the standpoint of title, it is obvious that without field survey information, you have no verification of the angular relation between the line bearing "North 38° 26′ East" and Ferris' south line of "South 89° 36′ East." If there was a survey and you knew the angle and distance and the bearing derived therefrom, you would do better to write the description thus:

> Beginning at the southeast corner of Lot 12 of Livingston's Tract as shown on a map recorded in Book 36, Page 24, records of said county; thence North 0° 48′ West 128.00 feet along the east line of said lot; thence North 38° 26′ East 116.74 feet to the south line of the land described in deed to William Ferris recorded in Book 675, Page 58 of Deeds; thence South 89° 52′ East 199.08 feet

> (cited in said deed as South 89° 36′ East 198.32 feet) along said south line to a 1″ iron rod; thence South 0° 08′ West 218.42 feet to the easterly prolongation of the southerly line of said Livingston's Tract; thence South 89° 52′ 14″ West 269.35 feet along said prolongation to the point of beginning.

In this way, with both the survey facts and lines of title to make a good legal description, there is no need to change the base of bearings; just relate the deed bearing to the one derived from the survey.

Another example reads thus:

> Beginning at the intersection of the west line of Howard Street with the north line of Bee Street; thence North 88° 57′ West 362.72 feet along the said north line (shown as North 89° 16′ West on the map of Keefer's Subdivision recorded in Book 72, Page 36, of Maps, records of said county) to the southeast corner of said Keefer's Subdivision; thence along the east line thereof North 1° 02′ West . . . . .

From these facts as stated, you do not know whether the angle between the east line of Keefer's Subdivision and the north line of Bee Street is based on the first bearing given or the bearing shown on the map. This type of writing is clearly ambiguous and the derivation of its meaning is impossible without a field survey. If, however, the above description started with reference to the map of Keefer's Subdivision, with or without the alternate bearing for Bee Street given parenthetically, you could rightfully assume and use the bearings on the subdivision map for developing the description unless or until another base of bearing was introduced.

There is no limit to the number of different base of bearings you may have in a single description depending upon the number of ties. This does not mean, however, that every tie necessarily incorporates a new base of bearing. This is an item you must watch very carefully, both in deciphering descriptions already on record, as well as in writing new descriptions.

In some cases, an *assumed* bearing may be assigned to the initial line. This is not generally desirable because correlation with subsequent lines related to matters of record becomes indeterminate until a field survey produces the actual relationship.

Sometimes the transposition of a description into map form is questionable until the changes in the basis of bearings are taken into consideration.

Azimuths

Some descriptions are written on the basis of azimuths, which means that the lines are developed from horizontal angles measured clockwise from either true north or true south, and they may be expressed as any angle between 0° and 360°. In the United States of America, there are two categories; the civilian azimuth, which is measured from true south, and the military azimuth, which is measured from true north. If it is unqualified, it is assumed to be civilian-oriented; however, if you are creating a description using azimuths, it would be well to designate the base so there is no misunderstanding. One rare exception of this is the use of an assumed azimuth based on some stated line other than north or south; this is not recommended.

When the azimuth is used in lieu of bearings, one form would read thus:

> "Thence on azimuth 52° 18′ (civilian) a distance of 328.42 feet; thence . . . . ."

The reference to "civilian" need be recited only once in the beginning. On this basis, the angle, being measured clockwise from the south, would be the same as the bearing "South 52° 18′ West." Likewise, azimuth 241° 53′ would be also the bearing North 61° 53′ East.

Another form would be:

> "Thence, on the basis of a line true south and passing through the westerly terminus of that course cited in deed to H. W. Brown as 'South 79° 48′ West,' on azimuth 211° 38′ a distance of 89.67 feet."

In each instance, each and every azimuth given would be based on a strictly true south. By the second form, all azimuths would be based on the initially stated line. There would be no material difference in small areas, but where the lines traverse far enough to get beyond the condition where meridians are not parallel within the accuracy of the instant measurements, you would have variations demanding consideration.

If an azimuth is based on magnetic north (or south), it should also state that fact and then if it must be related to true north, the amount of variation which existed at the date and place of the observation in the United States, plus the amount of change per year, must be known and that differential applied to the reading for rectification. This information is available from the National Oceanic Services

Administration, NOSA, (formerly United States Coast and Geodetic Survey) and is shown on an isogonic chart which they produce every five years. The amount of deflection between true north and the pointing of the magnetic needle varies irregularly across the country, except in one narrow band which registers no deflection. When the compass is west of this zero band, the north-pointing end of the compass needle deflects to the east and is said to have a positive (or plus) magnetic declination. Obversely, when it is east of zero, it deflects to the west and is said to have a negative (or minus) magnetic declination. The isogonic map also shows the annual rate of change for the different areas across the country.

An example of negative declination is shown in Figure 4 where the compass needle points 12° 18′ west of true north. On a "military" basis, the true azimuth for the line bearing S 46° 14′ W is 226° 14′ and on the magnetic azimuth, the angle is 238° 32′. A positive declination is shown in Figure 5 where the compass needle points 19° 10′ east of North. Again on the "military" basis, the true azimuth for the line bearing S41° 07′ E is 138° 53′ and on the magnetic azimuth, the angle is 119° 43′.

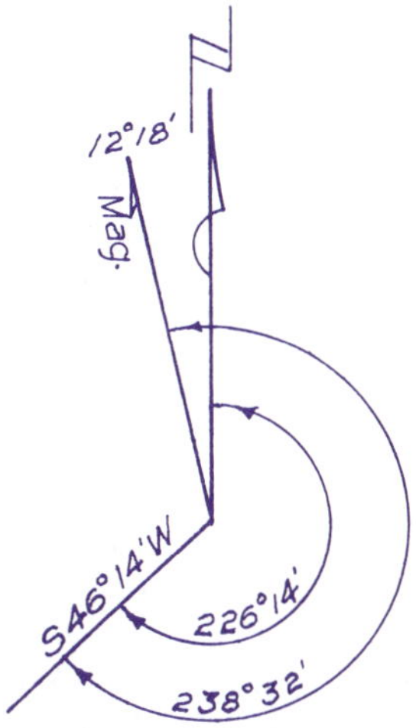

Fig. 4

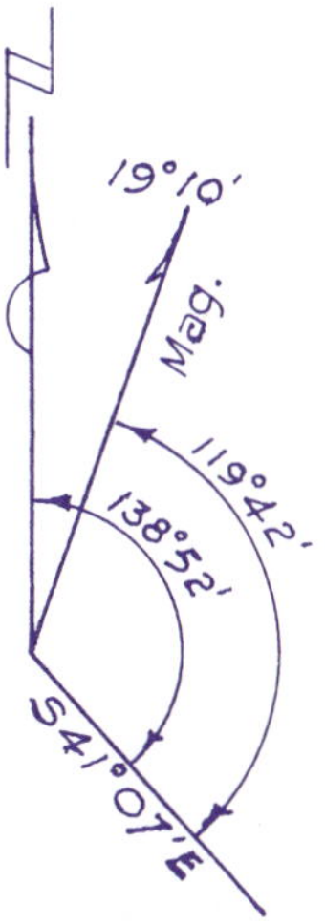

Fig. 5

If you find a description that states:

"A military azimuth of 92° 45′ based on magnetic declination of 14° 30′ east . . . ."
(See Figure 6.)

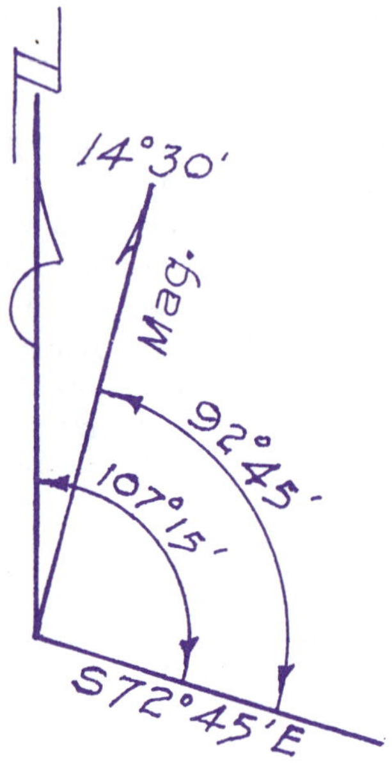

Fig. 6

this would mean that the 92° 45′ will be added to the 14° 30′ east to give a total angular measurement from true north (astronomical) of 107° 15′. This in turn calculates to a bearing of south 72° 45′ east.

To convert "military" azimuth to "civilian," add 180° to north azimuth angles between 0° and 180° and subtract 180° from north azimuth angles between 180° and 360° or vice versa from "civilian" to "military."

'ly Parts of an Area

The expression of norther*ly*, northeaster*ly*, etc., is general in compass direction but the distance measurement to which it is attached remains definite. Each " 'ly" sector of the compass embraces 45° which is split 22½° each side of the named direction. For example, in the drawing, Figure 7, *northwesterly* embraces 45° between the bearings N 22½° W and N 67½° W. Likewise, *southwesterly* embraces 45° between the bearings S 22½° W and S 67½° W. Also, *westerly* embraces 45° between N 67½° W and S 67½° W. *Northerly* embraces 45° between N 22½° W and N 22½° E and *southerly* embraces 45° between S 22½° W and S 22½° E. The same form of division applies to the east half of the circle.

The limits expressed above, however, are not rigid. In other words, if a line calculates to a bearing of N 68° 10′ W and you want to use an 'ly form of direction instead of the bearing, "northwesterly" would be acceptable, even though it was actually in the realm of "westerly."

The same is true for expressing the direction of the concavity of a curve which is based on the bearing of a line passing through the mid-point of the arc toward the center of that circle. If it is near the bearing between two 'ly directions, either one can be used or both can be expressed. More on this in the section on Curves.

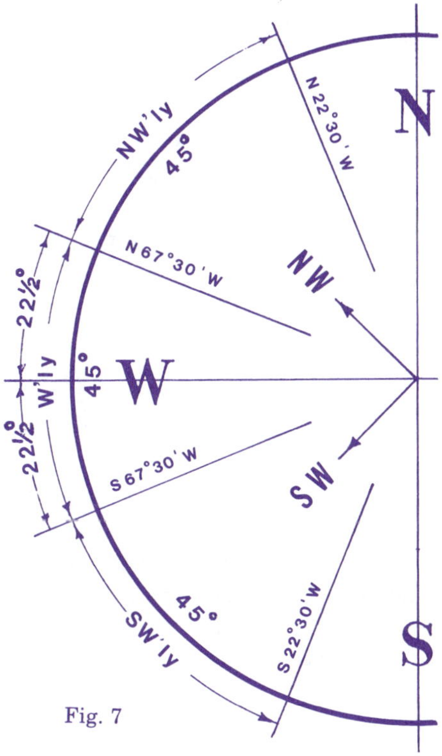

Fig. 7

When such a direction is applied to the segregation of an area in the expression, "the north (or northerly) 50.00 feet of Lot 6," it means that the south line of that 50.00 feet is parallel with and measured from the north line of that lot. As in Figure 8 herewith, "the east (or

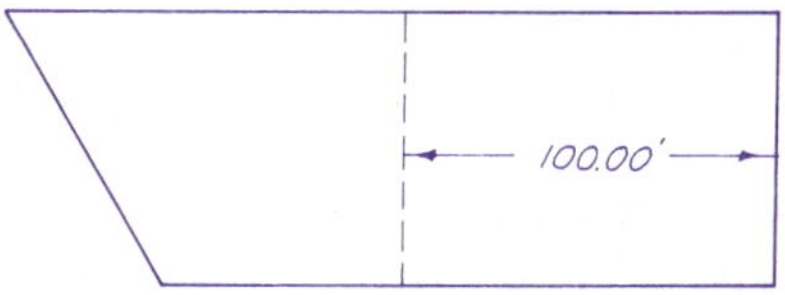

Fig. 8

easterly) 100.00 feet of Lot___" is an area bounded by a line parallel with the east line of the lot and set at 100.00 feet west measured at right angles from the east line of the lot.

In the case of irregular shaped parcels, Figure 9, remember that the expression, "the easterly (or east) 'x' feet of Lot___" means that the distance is measured at right angles, or radially, to or from (in this case) the east line, unless otherwise qualified. The courts have held that the grantee is to be favored and in order to give the maximum width, it must be measured at right angles or radially.

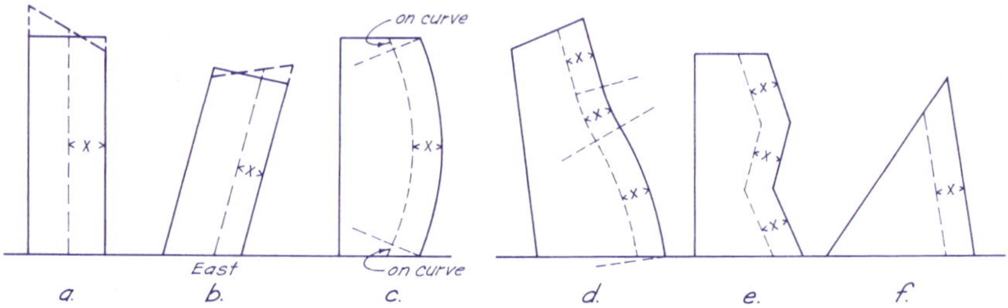

Fig. 9

If one were to describe an area as "the easterly (or east) 'x' feet of ___" and the parcel is the shape of, or similar to, those shown in Figures 10a and 10b, it would be indeterminate because by the rule of parallel lines for this type of description, there is no east line by which to apply parallelism.

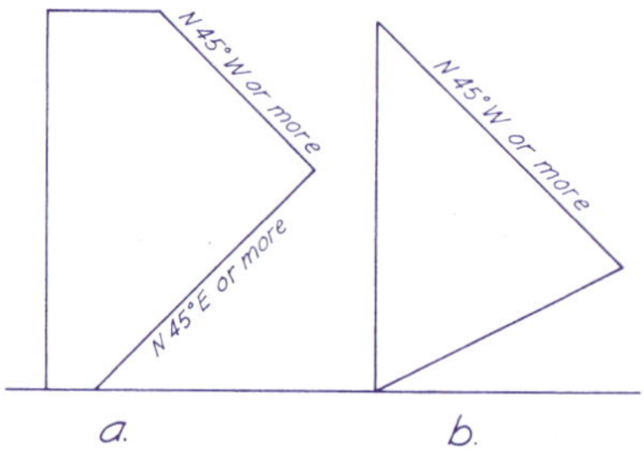

Fig. 10

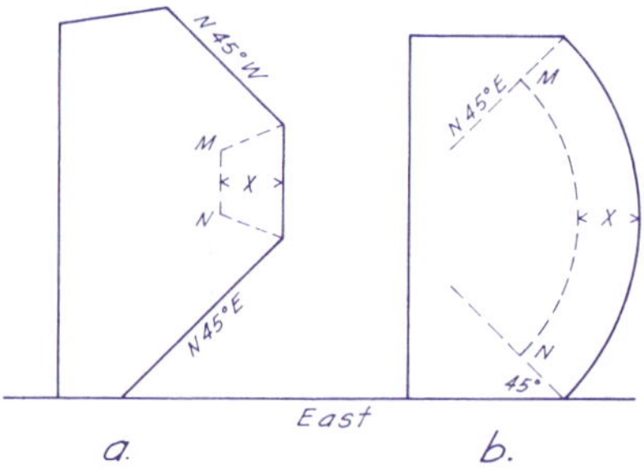

Fig. 11

With shapes of parcels as shown in Figures 11a and 11b, the statement, "the easterly 'x' feet of Lot___" is effective only between M and N. In the case of the curve, the limits are the 45° radial lines, whereas with the diagonal boundaries joining the east line, the lines bisecting the interior angles determine the limits.

If it is desired to describe, and cover with certainty, a strip of 'x' feet in width all the way from the north line to the south line, as shown in

Figure 12a, use the form "the northeasterly, easterly and southeasterly 'x' feet of Lot____." However, with the curve as illustrated in Figure 12b, the above form will not suffice; use this pattern instead: "That portion of Lot____ lying easterly of a curve concentric with and 'x' feet westerly from the easterly line of said lot (or, easterly line thereof) and its northwesterly and southwesterly continuation." (See distinctive meaning under *Prolongation* in Words and Phrases in Chapter 3.) There is no need to recite the limits of the north and south lines of the lot because the phrase, "that portion of Lot____" confines the strip and the continuation of the curve beyond the 45° radial lines to the lot lines.

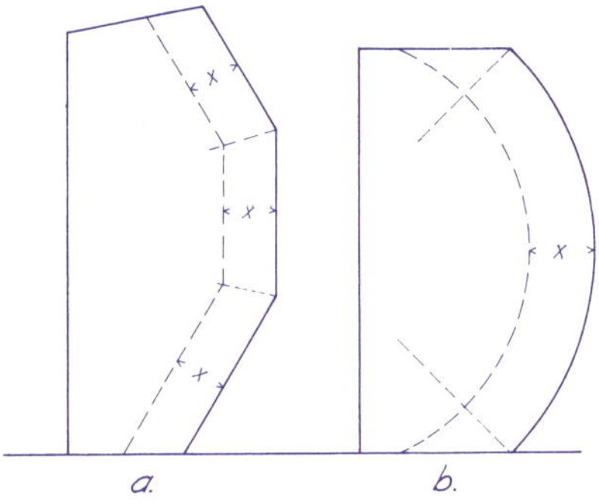

Fig. 12

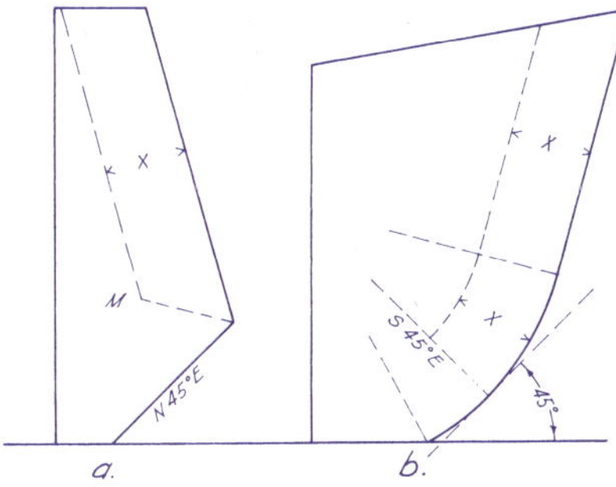

Fig. 13

In the case of parcels shaped as shown in Figures 13a and 13b, "the easterly (or east) 'x' feet of Lot____" is effective only north of M. As with the previous example, if the entire length of the lot is to be covered with the strip, then for Figure 13a, use the form "the easterly and southeasterly 'x' feet of Lot____." Again, with the curve situation in Figure 13b, as completed in Figure 14a, use: "That portion of Lot____ lying easterly of a line parallel with and 'x' feet westerly from the tangential easterly line of said lot (or, easterly line thereof) and easterly of a curve concentric with and 'x' feet westerly from the curved easterly line of said lot (or, curved easterly line thereof) and its southwesterly continuation."

If the southwesterly segment of the northwesterly line of the strip in Figure 14a is to be a tangent to the curve, instead of a continuation of the curve, then that northwesterly line should preferably be described completely and separately from its beginning on the north line of the parcel with all the calculations made to carry it through the curve and its tangent to the south line.

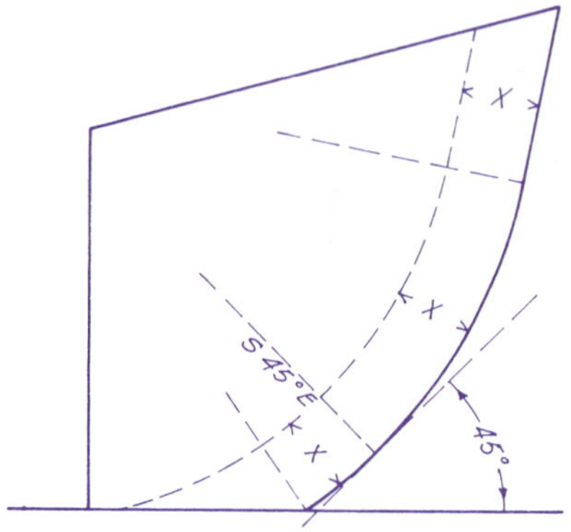

Fig. 14a

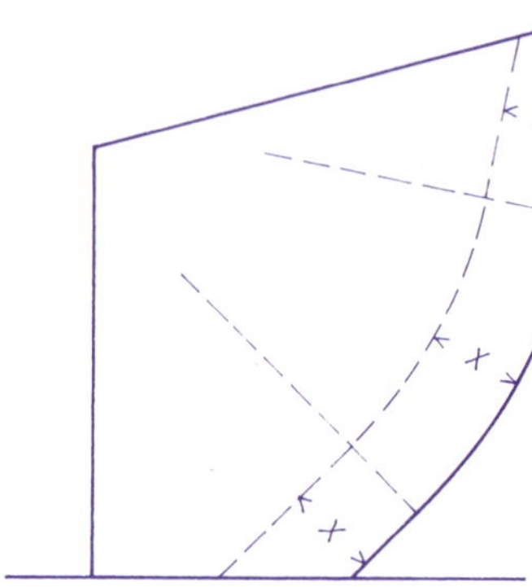

Fig. 14b

If the parcel is shaped as shown in Figure 14b with the east line having a tangent segment between the curve and the south line, you can use the simple form, "the easterly and southeasterly 'x' feet of Lot___."

Where is "the westerly '30' feet of Lot 4" at the southwest corner with a curved corner as shown in Figure 15? Any portion southerly of the radial line at the beginning of the northerly end of the curve is indeterminate. If it is desired to have a straight line cut-off, the description could read, "That portion of Lot 4 lying westerly of a line parallel with and 30 feet easterly of the tangential portion of the westerly line of said lot (or, easterly of the tangential portion of the westerly line thereof)." This is another example of the rule given for the word *Parallel* in Words and Phrases under Chapter 3 where, in this case, the easterly line will automatically be prolonged to the southerly line without reciting, "and the prolongation thereof."

However, if the easterly line of said strip is to be other than entirely straight, it should be described with every part in its entirety from boundary to boundary of the lot.

Another example of parting a corner lot with a curved corner is shown in Figure 16. Again, "The west 'w' feet of Lot___" is indeterminate but "The east 'x' feet of Lot___" is sufficient; therefore, a safe way to convey the westerly portion is to use the description form, "Lot___

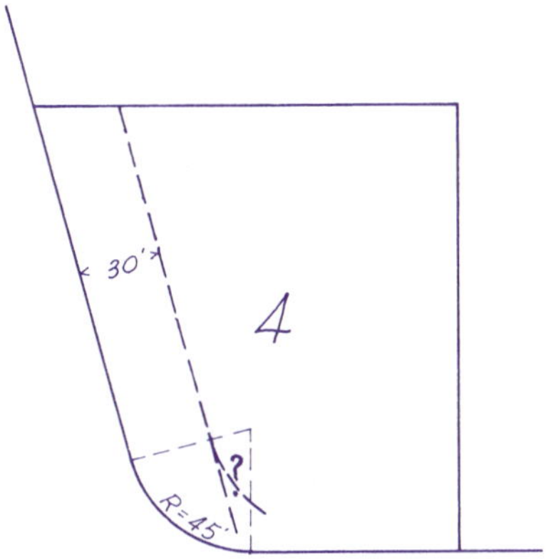

Fig. 15

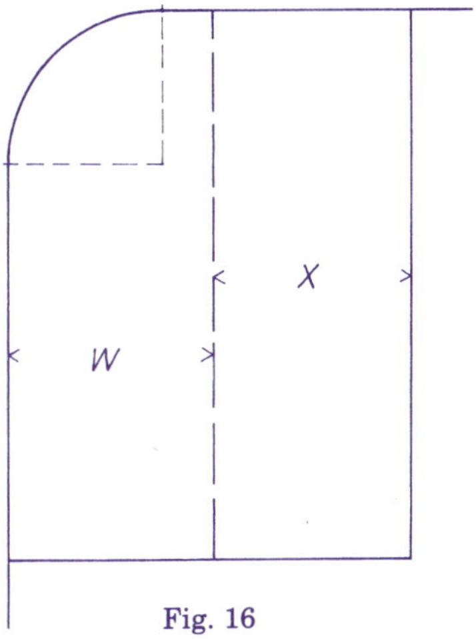

Fig. 16

in Tract No.___ in the City of___ etc. EXCEPT the east 'x' thereof (or, of said Lot)."

With a stair-step parcel as in Figure 17, "the easterly (or east) 'x' feet of Lot___" is indeterminate because the statement could apply to either the north or south portion. The southerly portion could be described as, "The most easterly 'x' feet of Lot___." The northerly portion would be better described as "The easterly 'x' feet of the northerly 'y' feet of Lot___."

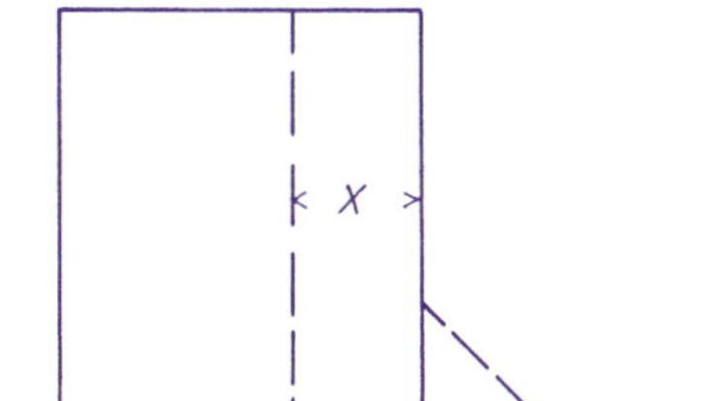

Fig. 17

The previous examples have generally treated just one part of a lot or parcel. Let us consider now three methods of measuring such a part and the result of their applications. In Figure 18, three different lines are shown in relation to the westerly line of a parcel. The line *a* labeled "W 50′" is by definition parallel with and measured at right angles from the westerly line. The line *b* labeled "W 50′ measured on north line" is closer to the westerly line and parallel with it. The line *c* labeled "W 50′ measured along the north and south lines" is not parallel with the westerly line. Although this has been also expressed as "W 50′ measured front and rear," the form is not recommended as explained in Words and Phrases under Chapter 3.

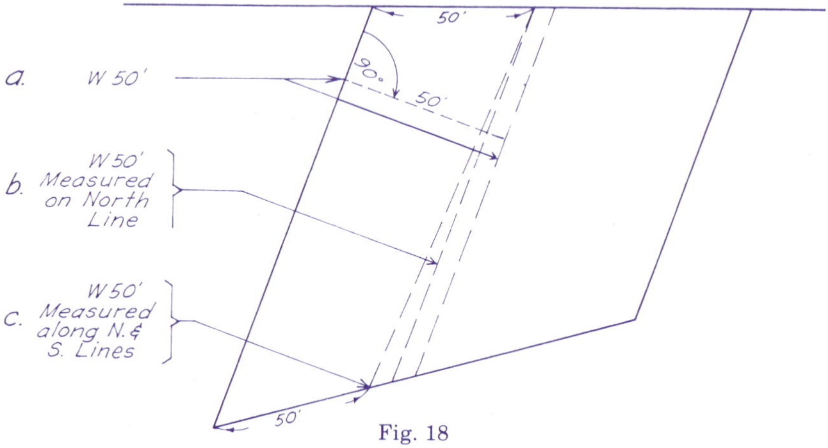

Fig. 18

The format used for the easterly portion shown in Figure 8 cannot be used to describe the westerly portion as shown in Figure 19 because, by the rule, the easterly line would have to be parallel with the westerly line and, using the remaining distance of 70 feet, would encroach into "the east 100 feet of Lot___." This is so whether you use "the west 70 feet . . . ." or "the west 70 feet measured along the north line." If you used some lesser distance, it would leave a gap.

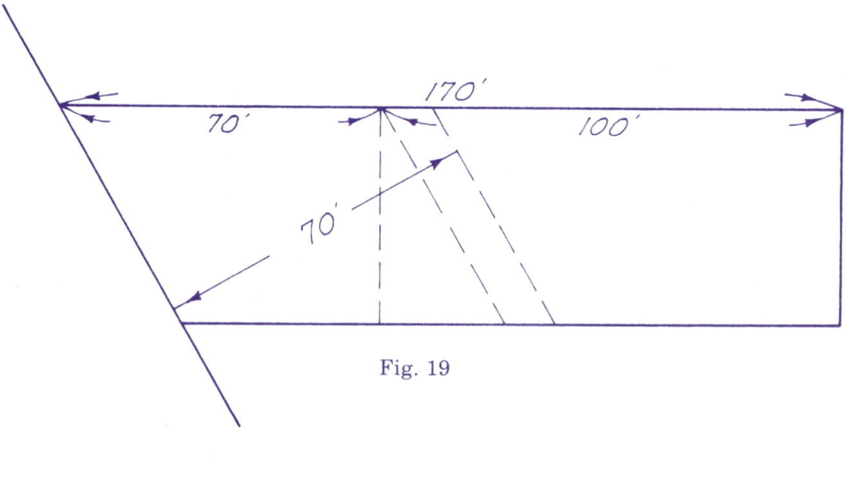

Fig. 19

The correct way to cover the westerly trapezoidal portion is to use the form, "Lot___ in Tract No.___ in the city of.......... EXCEPT the east 100.00 feet thereof."

With the obverse situation calling for "the west 70 feet measured along the north and south lines" as shown in Figure 20, you cannot describe the remainder here either using the 100 feet. Again, it must be covered by using the form, "Lot___ in Tract No.___ in the city .... EXCEPT the west 70 feet measured along the north and south lines thereof."

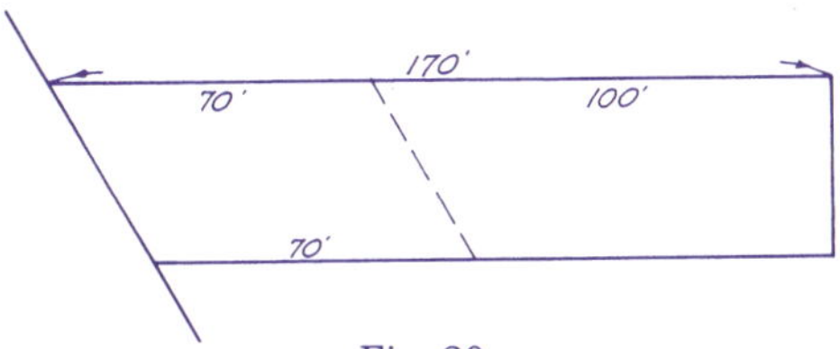

Fig. 20

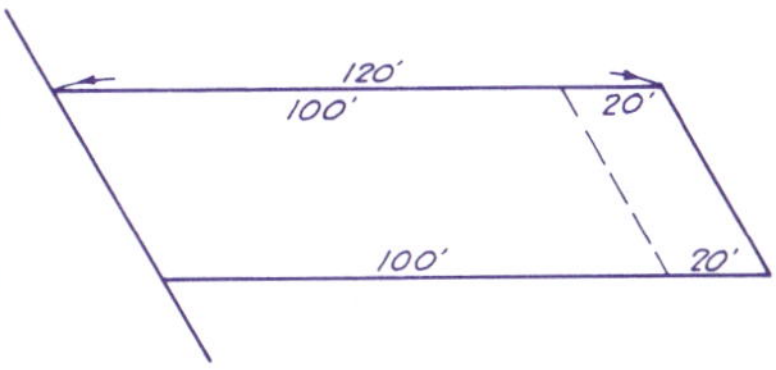

Fig. 21

In the case of a parallelogram lot as in Figure 21 where it is desired to dispose of "the west 100 feet" and "the east 20 feet, measured along the north and south lines," *DO NOT* use *both* measurements. It might appear to be acceptable on the face of the map, but, if there is any difference whatsoever on the ground, you will have either an overlap or a gap. The totally safe way is to use one distance only (usually the first sale out) and use the "EXCEPT" form of description for the remainder. By the latter method, you can have no gap or overlap.

With the parallelogram lot shown in Figure 22, the assumption of describing "the west 100 feet" and "the east 40 feet" with 140 feet shown on the north line is obviously in error. Again, the safe way is to use the distance decided upon for the first sale and then use the "EXCEPT" form of description for the remainder.

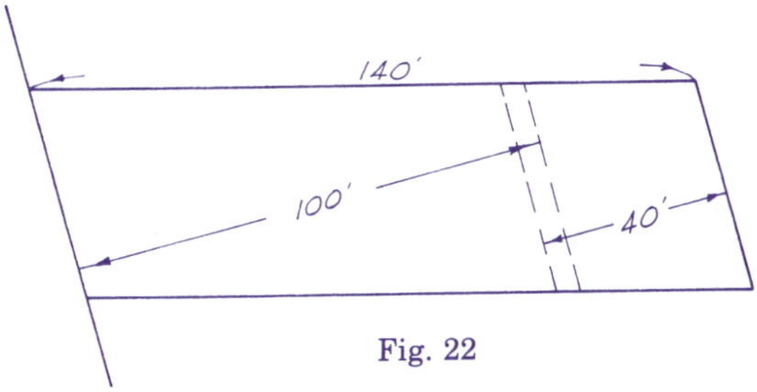

Fig. 22

4.18

Referring back to Figure 18, let us take the two applications *a* and *b*

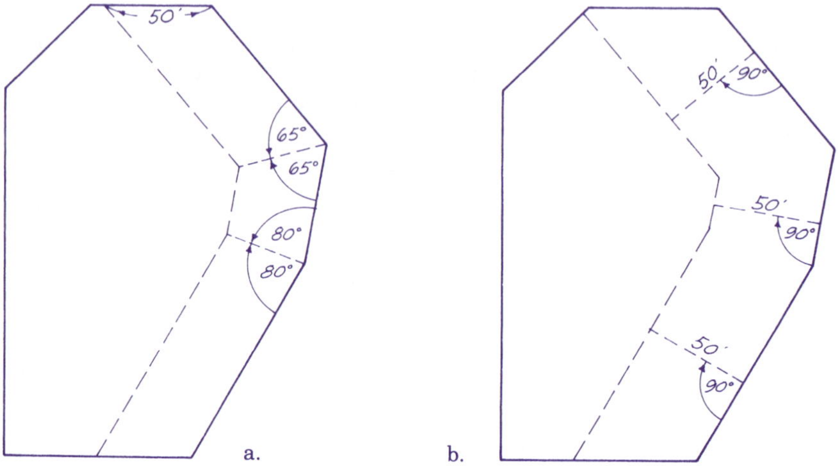

Fig. 23

and their difference when used on a multi-angular side of a parcel as shown in Figures 23a and b. "The easterly 50 feet measured along the north line" is the narrower of the two strips. "The easterly 50 feet" being measured at right angles from each segment is of course wider.

So far, these examples have dealt with one part of a parcel using a single direction and distance. It is possible to expand this format into multiple parts using two directions and two distances, three directions and three distances, four directions and four distances, etc., etc.

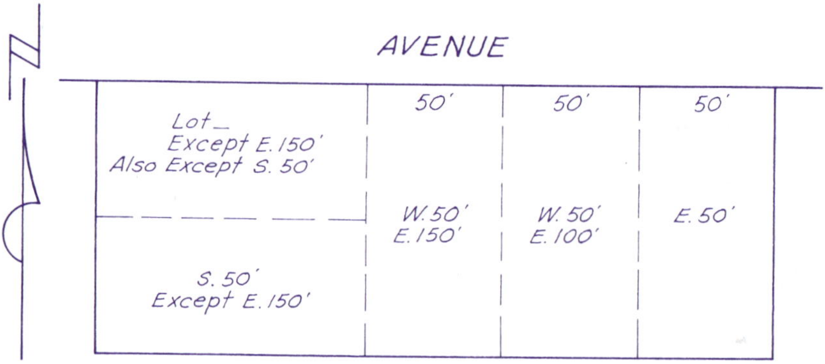

Fig. 24

For example, in Figure 24, the second 50-foot parcel would be "the west 50 feet of the east 100 feet of Lot___." The third one would be "the west 50 feet of the east 150 feet of Lot___." In order to have no conflict, the southwest parcel would be "the south 50 feet of Lot___ . . . . EXCEPT the east 150 feet thereof." Having established the first four parcels with definite distances, the last one in the northwest corner should take whatever remains in the Lot by exceptions: "Lot___ . . . . EXCEPT the east 150 feet thereof. ALSO EXCEPT the south 50 feet thereof."

Note that the second "EXCEPT" is preceded by the word "ALSO." This is a necessary form so that each exception stands independently.

In Figure 25, you have "Parcel 5 as per map recorded in book 21, page 32 of Parcel Maps. EXCEPT THE EAST 30 feet thereof. ALSO EXCEPT THE SOUTH 15 feet thereof." This type of double exception is acceptable and there is no problem with the overlap.

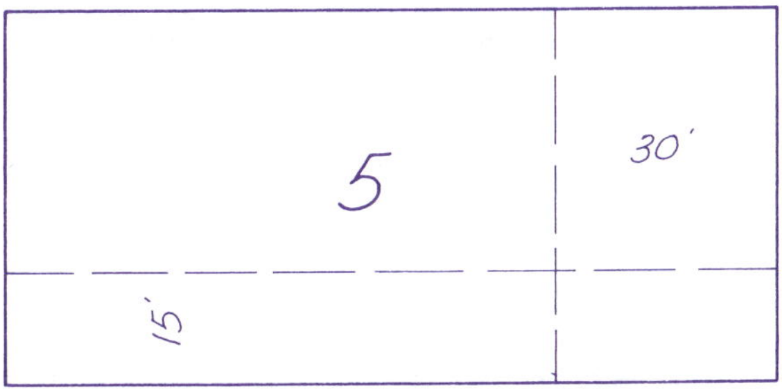

Fig. 25

An example of a triple call is shown in Figure 26 where a 30′ x 60′ piece is taken out of the lot after a 20′ widening of the street, thus: "The south 30 feet of the east 60 feet of the west 80 feet of Lot___ in the City of . . . ."

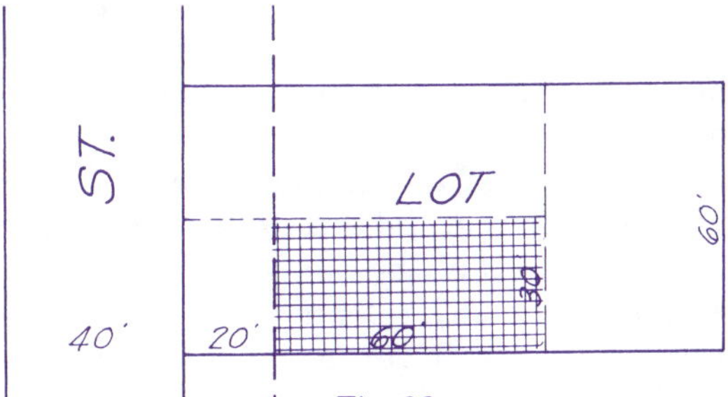

Fig. 26

A quadruple call is illustrated in Figure 27 and the description would read: "The south 60 feet of the north 300 feet of the east 120 feet of the west 450 feet of the southwest quarter of the southwest quarter of the southwest quarter of Section . . . ."

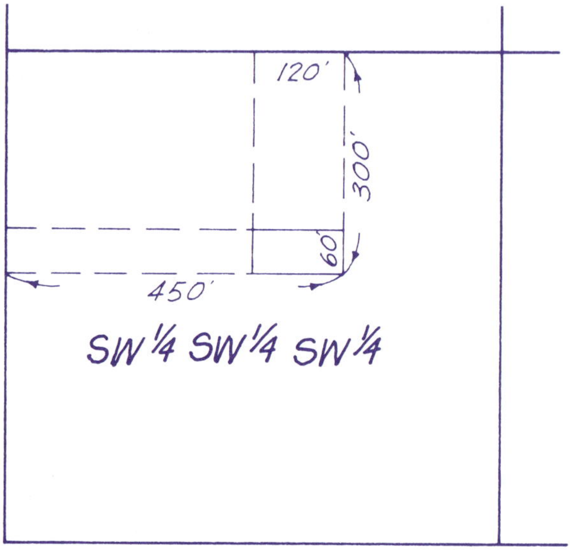

Fig. 27

The use of this short form 'ly type of description is more direct and easier than the sometimes cumbersome metes and bounds writings.

Curves in Surveys and Descriptions

Curves become an integral part of highways, railroads, canals, right-of-ways, subdivisions, and property boundaries when the surface conditions of the land or the esthetic design of subdivisions require such configurations; therefore, one must know the methods of creation of curves, their interpretation and proper delineation for correct descriptive application. In this discussion, curves will be treated only in the horizontal plane.

A CURVE is a segment of the circumference of a circle. Its end limits are the radii of the circle and the angle between those two radii is the central angle or delta of the curve.

The correlated parts, known as elements of a curve, are: radius (R), length (L), central angle or delta (Δ), semi-tangent (usually just called tangent) (T), and chord (C). See Figure 28.

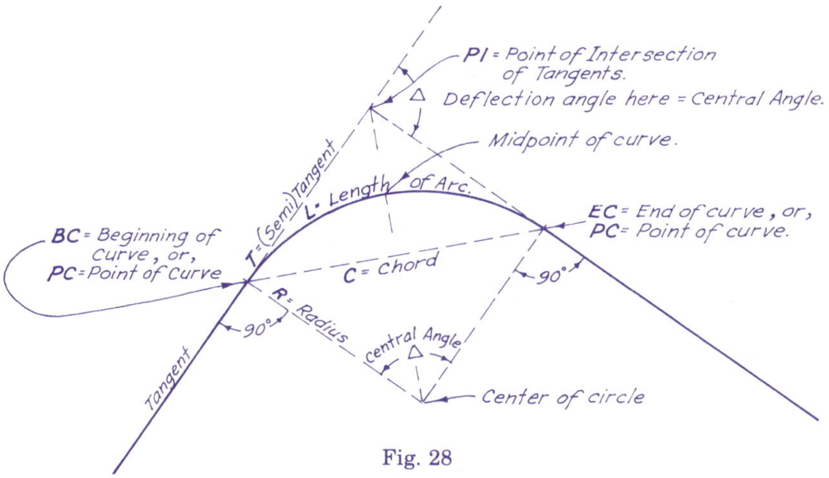

Fig. 28

-a- RADIUS is the line between the center of the circle and its circumference.
-b- The LENGTH of arc is the distance along the curve.
-c- The CENTRAL ANGLE or DELTA is the angle subtended by the radii.
-d- The SEMI-TANGENT (generally just called TANGENT) is the extension of the straight line (sometimes called the tangent line) which precedes or follows the curve to its intersection with the other semi-tangent.
-e- The CHORD is that segment of a straight line which is intersected by two points on a curve, or, in other words, the straight line distance between the two ends of a segment of arc.

The illustration in Figure 29 shows the application of the following qualifications of curves:

-a- A *simple curve* is a single segment of arc.
-b- *Compound curves* are a group of two or more segments of arcs having different lengths of radii on the same side of the curves, common radial lines and common tangents.
-c- A *spiral curve* is a collective group of multiple compound curves having radii of successively decreasing or increasing lengths.
-d- A *reverse curve* is a segment of arc of which the center of its circle is on the opposite side of the adjoining curve and of which one of its radial lines is the prolongation of a radial line of the opposite curve.

-e- *Tangent curves* have a common radius or a prolongation of each other's radius (though they may be of different lengths); therefore, compound curves and reverse curves are by definition tangent.

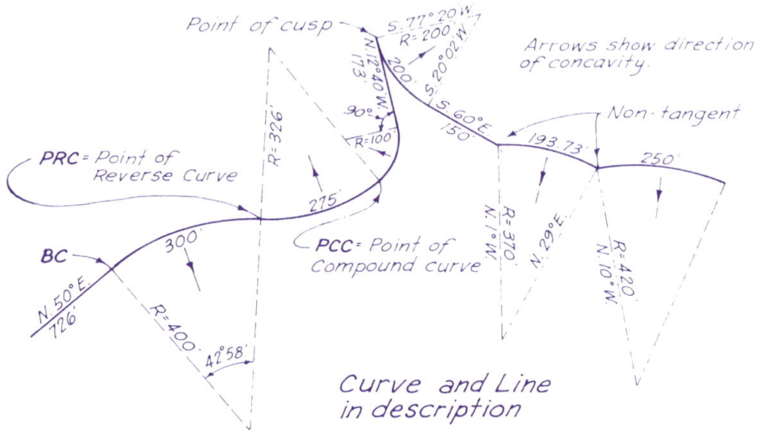

Fig. 29

-f- *Tangency* of a curve to a line or curve means that the radius line is perpendicular to that line or the tangent of the curve at that point of curve.

-g- *Concavity* relates to the enclosed side of a curve toward the center of the circle. The general direction given for the concavity is that of the radial line from the midpoint of the arc toward the center.

-h- *Convexity* (seldom used) relates to the outside face of a curve away from the center.

-i- *Non-tangent* relates to the fact that the beginning of the curve is not tangent to the preceding line or curve. This condition is also known as a *broken-back* curve.

-j- A *cusp* is the meeting of two curves or a curve and a straight line having the same general direction. The point of a crescent moon is a cusp.

-k- The *direction* given along the curve is a general one (NE'ly, S'ly, etc.) for the total arc; however, if the total angle is more than about 120°, two or more directions may be combined, such as W'ly and NW'ly. See Figure 30. In the case of a cul-de-sac as in Figure 31, three directions would be desirable, i.e., "thence 286.74 feet northerly, easterly and southerly, along a curve concave to the south having a radius of 47.00 feet."

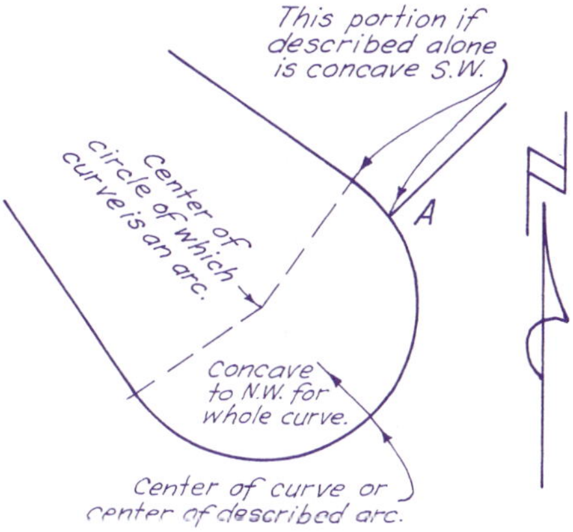

Fig. 30

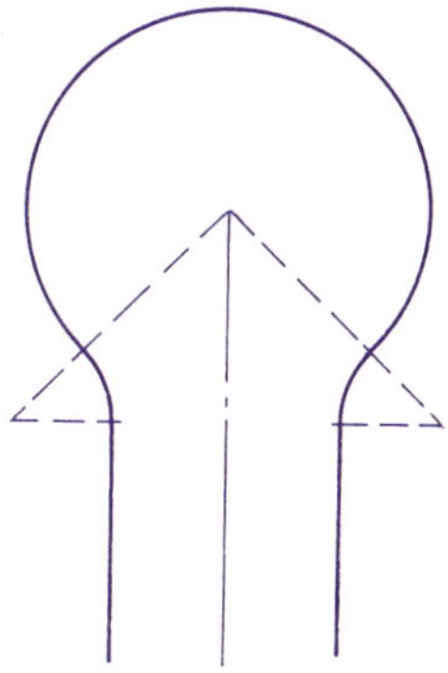

Fig. 31

Other points related to the curve shown in Figure 28 and used in descriptions are:

-a- Beginning of curve (BC) which is the point in the description where the curve begins regardless of whether it is tangent or non-tangent. It is sometimes called a point of curve (PC).

-b- Beginning of compound curve (BCC) which is the point in the description where one curve ends and an adjoining tangent curve having a common radius begins. It is sometimes called a point of compound curve (PCC).

-c- Beginning of reverse curve (BRC) which is the point in the description where one curve ends and an adjoining tangent curve concave in the opposite direction begins. It is sometimes called a point of reverse curve (PRC).

-d- Midpoint of curve is the halfway point along the arc between the beginning and end of the curve. It lies on the bisection of the central angle. This is very seldom used.

-e- Point of intersection of tangents (PI) is the point at which the prolongation of tangents of a curve, one passing through each end of the curve, intersect, but not possible when Δ is more than $179°59'59''$.

-f- End of curve (EC) which is the point in the description where the curve ends. It is also sometimes called a point of curve (PC). This statement is seldom used.

-g- The center of a circle is the point from which the radial line bears.

The most commonly used elements are the radius and length. Sometimes the central angle is added. In some instances, the bearing and length of the chord is given. The third element can be used for a check on the others unless it creates conflict, in which case, extrinsic evidence may be necessary for a solution.

Before proceeding with the multiple curve calls and curve description writing, let us consider the problems involved by the use of the words "right" or "left." The use of the words "right" or "left" to connote one side or the other is not fully satisfactory. In the early days, the courses of a river would be described as following the left or right bank going downstream. Attempts were made to adapt this to other uses, but inherent variables made it uncertain. See Figure 32. A compass direction or a positive bearing is much to be preferred.

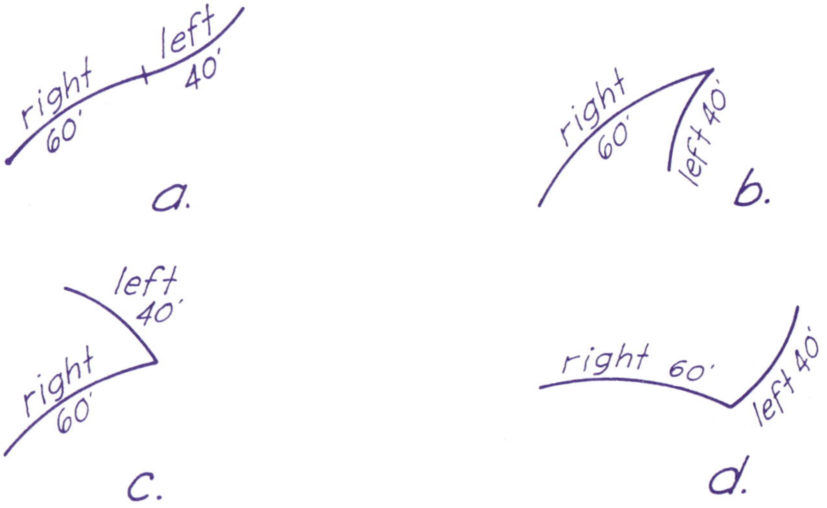

Fig. 32

If your description reads, "Thence along a curve to the right 60 feet, thence along a curve to the left 40 feet . . . ," then examples a, b, c and d in Figure 32 would all be apropos but certainly not finite to the particular condition at hand.

Consider a more certain way by the use of compass directions, still referring to Figure 32.

a. "Thence northeasterly 60 feet along a curve concave to the southeast having a radius of 100 feet to the beginning of a reverse curve concave to the northwest having a radius of 60 feet; thence northeasterly 40 feet along said curve."

Unless otherwise stated, both a reverse and a compound curve are tangent to their previous curves so this is compatible with Figure 32a.

Continuing with examples b, c, and d:

b. "Thence easterly 60 feet along a curve concave to the south having a radius of 60 feet to the cusp of a curve concave to the southeast having a radius of 50 feet; thence southwesterly 40 feet along said curve."

And: c. "Thence northeasterly 60 feet along a curve concave to the southeast having a radius of 200 feet to the cusp of a curve concave to the southwest having a radius of 70 feet; thence northwesterly 40 feet along said curve."

Also: d. "Thence easterly 60 feet along a curve concave to the south having a radius of 200 feet to the beginning of a non-tangent curve concave to the northwest having a radius of 75 feet; thence northeasterly 40 feet along said curve."

In each of these described examples, there is no question about the general position of each of the curves, their relation to each other, and their direction of travel. However, this is still not finite enough for surveying because each of the eight compass directions — north, northeast, east, southeast, etc. — include a spread of 45° or 22°30′ each side of the direction.

To make the above described curves finite enough to fix all their parts, the bearing of one of the radials or tangents of each curve should be included, thusly:

a. "Thence from a tangent line bearing N 38° 13′ E, northeasterly 60.00 feet along a curve concave to the southeast having a radius of 100.00 feet to the beginning of a reverse curve concave to the northwest having a radius of 60.00 feet; thence northeasterly 40.00 feet along said curve."

If the line previous to the curve had already been cited, it would be unnecessary to repeat the bearing because it would be properly assumed that the curve was tangent unless otherwise stated.

b. "Thence from a tangent line bearing N 29° 40′ E, easterly 60.00 feet along a curve concave to the south having a radius of 60.00 feet to the cusp of a curve concave to the southeast having a radius of 50.00 feet, to which point of cusp a radial of the last mentioned curve bears N 23° 15′ W; thence southwesterly 40.00 feet along said curve."

The bearings of radials (unless otherwise definitely stated) are expressed in a direction *from* the center of the circle to its circumference. The radial line stops at the circumference of its circle. If you wish to follow a radial line beyond that, you must use the expression, "along the prolongation of said radial line . . . . . ."

c. "Beginning at a point on a curve concave to the south having a radius of 200.00 feet, to which point a radial line bears N 47° 28′ W; thence easterly 60.00 feet along said curve to the cusp of a curve concave to the southwest, to which the center of the circle of said curve bears S 72° 06′ W 70.00 feet; thence northwesterly 40.00 feet along said curve."

The use of the bearing and distance to the center of the circle accomplishes two objectives in one statement: both the radial line and the radius are given.

d. "Beginning at a point N 2° 13′ E. 200.00 feet from the center of a circle; thence easterly 60.00 feet along the curve of said circle concave to the south to the beginning of a non-tangent curve concave to the northwest having a radius of 75.00 feet, to which beginning of curve a radial bears S 23° 14′ E; thence northeasterly 40.00 feet along said curve."

This time the combined radial line and radius were used to start the description. In all of the above cases, the bearing of the radial to the end of the first curve would be calculated.

With the bearings cited to control one point of each curve, it is possible to draw each pair of curves in exact relation to each other.

Curves are said to be concentric if they have a common center point and different radii and lie in the same plane. See Figure 33a.

Curves are parallel if they lie in different planes and the planes are parallel and their center points are in the same line through the planes. This would hold whether the radii were the same or different. See Figure 33b.

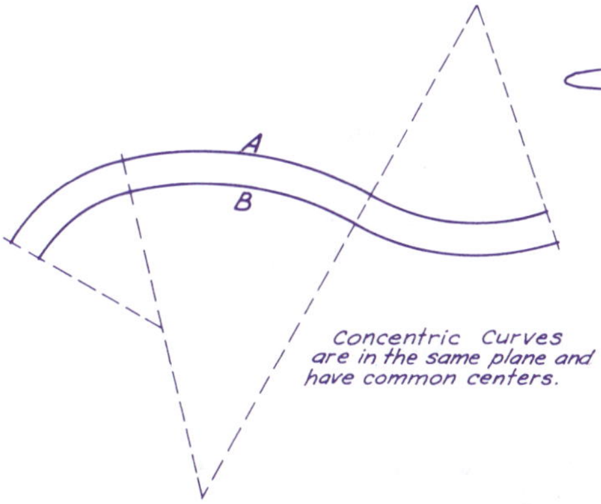

Fig. 33a

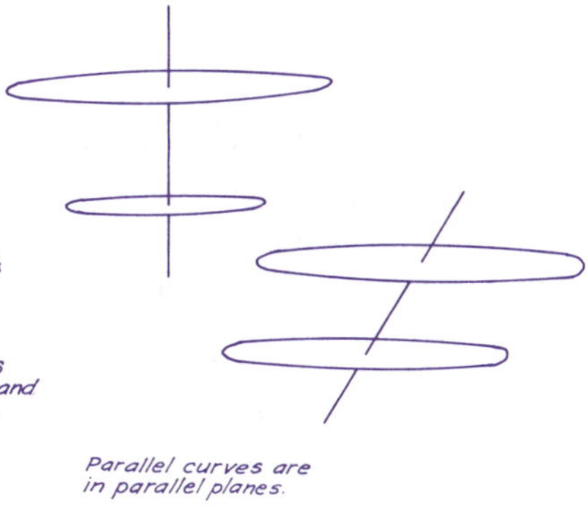

Fig. 33b

Two or more curves having the same radius but not common center points (whether in one or more planes) are considered to be equal but not concentric nor parallel. See Figure 34. Although corresponding points are at equal distances along parallel lines, the curves are still not accepted as parallel.

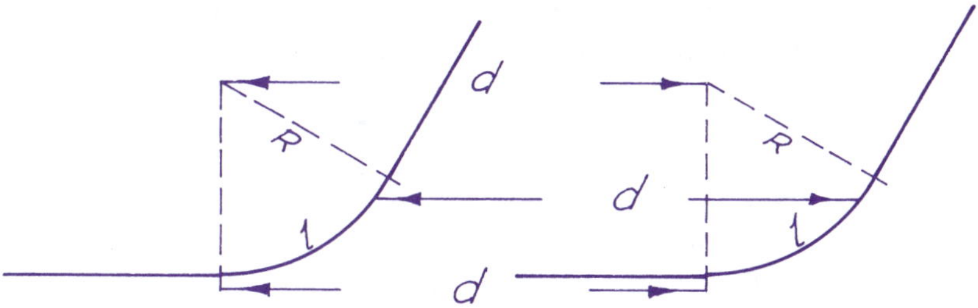

Fig. 34

You will note in the preceding descriptions of curves the use of the phrase, "along said curve." This expression is necessary in order to connote that the measurement of the line follows the arc of the curve and not across it on the chord.

Contrariwise, if the form were used, "thence easterly to a point on said curve 40.00 feet from the westerly terminus of said curve," or "easterly 40.00 feet to a point on said curve," this would connote that the distance is measured along the *chord* of that curve and not along the arc.

In writing legal descriptions, too much information is sometimes as detrimental as insufficient information. Great care must be taken in describing curves and their relationship to other lines because it is easily possible to use too many parts of a condition, some of which may not be compatible with certain other ones.

For example, this portion of a description follows curves to a line that is supposed to be both tangent and parallel with, and a certain distance from, an existing record right-of-way:

> ". . . . . to the beginning of a curve, tangent, concave westerly and having a radius of 480.00 feet; thence northerly and northwesterly along said curve through a central angle of 39° 38′ 26″ an arc distance of 332.09 feet to a point of reverse curvature (a radial through said point bears N 50° 37′ 34″ E), said point being the beginning of a curve, tangent, concave easterly and having a radius of 480.00 feet; thence northwesterly and northerly along said curve through a central angle of 39° 38′ 26″ an arc distance of 332.09 feet to a line tangent; said line being 41.50 feet easterly, measured at right angles from the easterly right of way line of the Edison Securities Company as described in deed recorded May 23, 1954 in book 2736, page 411, Official Records of said Orange County; thence N 0° 16′ 00″ E along a line parallel with and 41.50 feet easterly of said easterly right of way line . . ."

Note that all three conditions for both curves are cited and then qualified as terminating in a tangent that is also parallel with and 41.50 feet easterly from the right-of-way line. It is highly improbable that all of these conditions will remain intact; it would be remarkable if physical development of the curve concluded in a line that was both tangent to the curve *and* parallel with the right-of-way, *and* 41.50 feet from the right-of-way. See Figure 35.

Realizing that field conditions would most likely render this compatibility impossible, the creator of such a description would do well to decide first whether the condition of parallelism or tangency or fixed distance, or all three, should prevail and then use the form which would hold those conditions fixed and make the others variable and subject to the ultimate requirement.

In this case, if the line after the curve must be parallel, then only one condition of the curve should be given (whichever one is to control, i.e., radius, length, angle, or chord) and with no mention of its being tangent to said parallel line; however, this would probably result in a "broken-back" type of relationship between the curve and the following line. Otherwise, if the line at the end of the curve is to be tangent regardless of parallelism, then give two conditions of the curve and make no requirement of that line being parallel, or at any given distance from the right-of-way. If all three conditions in relation to the right-of-way must prevail, then the description must be written with variables to permit the length of the line or curve preceding the last curve to give in such a way that the last curve can be adapted to fit the combination.

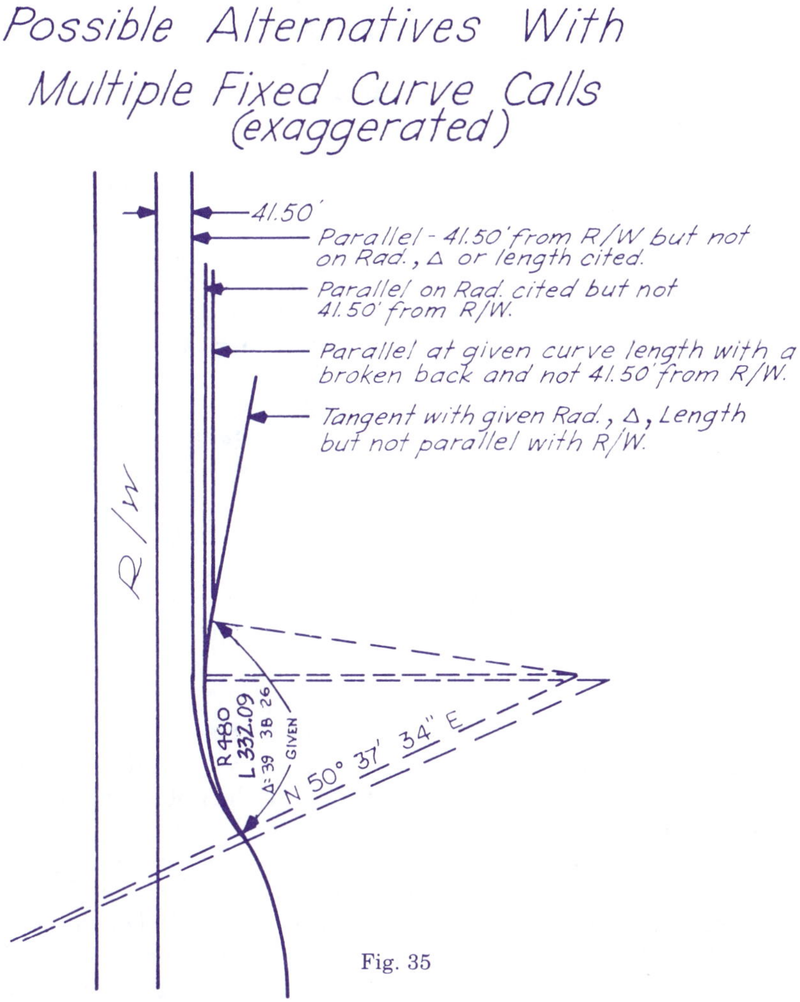

Fig. 35

For example:

.... to the beginning of a reverse curve (through which a radial bears N 50° 37′ 34″ E), said curve being concave northeasterly

> and tangent to a line parallel with and 41.50 feet easterly from the easterly right of way line; thence northwesterly and northerly along said curve to said tangent parallel line . . . . .

or

> . . . .; thence northerly and northwesterly 332 feet more or less to the beginning of a reverse curve having a radius of 480.00 feet, concave northeasterly and tangent to a line parallel with and 41.50 feet easterly from the easterly right of way line; thence northwesterly and northerly along said curve to said tangent parallel line . . . . .

Note, in the first case, that where the bearing of the radial line is fixed, all parts of the curve must be adaptable. In the second case, it is desired to hold the radius; therefore, both the radial bearing between the two curves and the length (which would also mean the angle) of the first curve must be left open to make all parts of both curves give in the combined demand of tangency and parallelism.

Fitting a curve tangent to two straight lines is a little simpler. Thus:

> . . . . .; thence N 48° 26′ 20″ W 124 feet more or less to the beginning of a curve concave northeasterly, having a radius of 480.00 feet and tangent to a line parallel with and 41.50 feet easterly from the easterly right of way line; thence northwesterly and northerly along said curve to its tangency with said parallel line . . . . .

Remember that redundancy in the form of excess information often causes impossible and/or unadjustable conditions.

Rehabilitation of Curves

In determining which elements should be held in describing or rehabilitating a curve when a field resurvey reports conflicting facts, one needs to know how curves are created. In the field location of the centerline of a road, either an existing one or a new one to be established, the general procedure is to locate the middle of the straight sectors and prolong their tangent lines to intersect and set those points of intersection (PI) for control. See Figure 36. The deflection angle at the PI is then measured, although sometimes the angular measurement is set first at an even figure and the tangent line moved to conform. If the ground conditions are unacceptable for setting a point, as perhaps it might be at PI #15 in the illustration, either the radius to the center of the circle, of which the arc will be tangent to the tangents, will be set off and measured or the chord will be established and measured for length and deflection angle. Of course, the deflections from the chord to both tangents must be equal.

After the deflection angle at the PI is measured (which by geometry equals the central angle), usually either the radius is assumed at an even number of feet or the semi-tangent is set at an even number of feet. In either case, the control is established and the remainder elements are calculated from them.

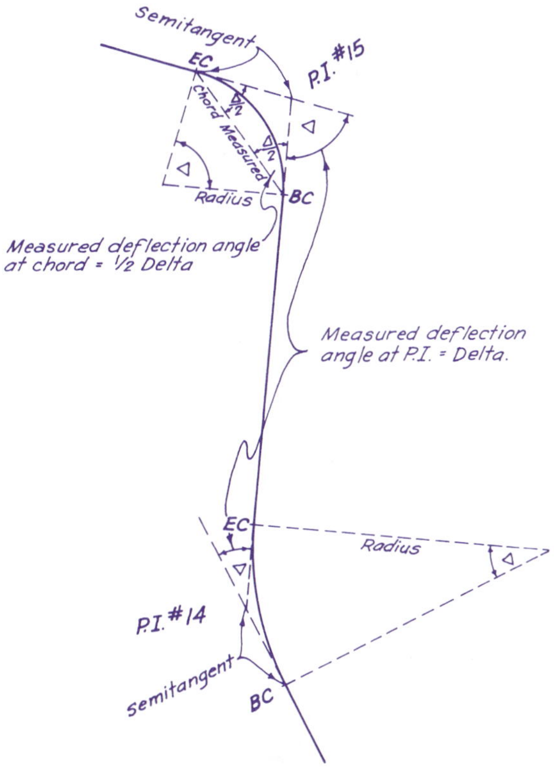

Fig. 36

It is because of the fact that two of these three elements (the central angle, semi-tangent or radius) are determined in the field to be the controls for the curve that you look to them first in the event of any discrepancies.

If, for example, the curve data is shown on a map as:

$$\begin{aligned} R &= 220.00' \\ \Delta &= 35^\circ\ 00'\ 00'' \\ L &= 134.39' \\ ST &= 69.37'' \end{aligned}$$

it is obvious that the creator set the angle and assigned the even foot radius.

If the field notes of a resurvey of this curve show:

$$\begin{aligned} R &= 222.31' \\ \Delta &= 34^\circ\ 57'\ 20'' \\ L &= 135.63' \\ ST &= 70.00' \end{aligned}$$

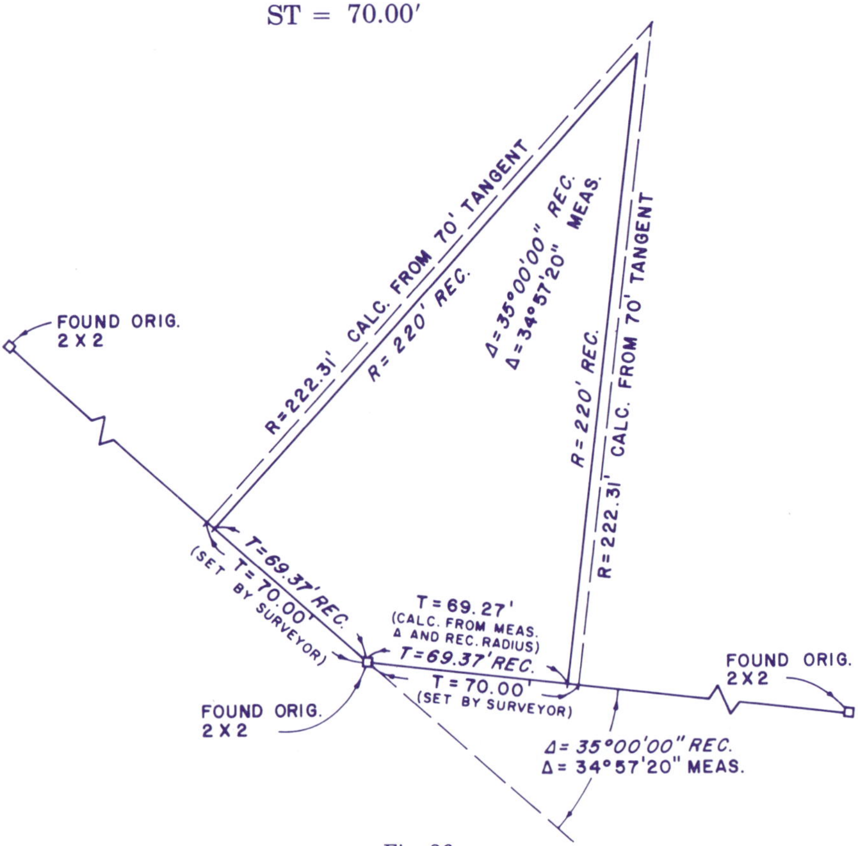

Fig. 36a

it would appear that the field angle was measured and found to be less, and that an even footage was arbitrarily assigned to the semi-tangent. See Figure 36a.

Which statements will hold? Do you accept them as submitted? Going back to the normal method of creation, it would appear that the even foot *radius* should hold and if the remeasured angle is less, as shown, then the semi-tangent and arc length would be calculated from that angle and the held radius, thus:

$$R = 220.00'$$
$$\Delta = 34^\circ\ 57'\ 20''$$
$$L = 134.22'$$
$$ST = 69.27'$$

The resurvey should not willfully change the element control as first established. On the other hand, neither should you write your description changing the surveyor's field note information without discussing the matter first. He may claim to have found one or more monuments upon which he based his decision; however, if one adheres to the well-established custom of following in the footsteps of the original surveyor, it would appear contradictory that he would show an even 220.00 foot radius on the map while setting stakes at an incompatible even 70.00 foot tangent.

Curve Format

The following description is for the purpose of giving you a format to use in the wording of various types of curves. See Figure 37.

That portion of Lot 6 in The Northwoods Tract in the City of ______________________________, County of ________________, State of ____________________, as per map recorded in Book ___, Page ___, of Maps in the office of the County Recorder of said County described as follows:

Beginning at a point on the west line of said lot S 01° 37′ 23″ E 198.00 feet along said line and its prolongation from the center line of NORTHWOODS AVENUE adjoining said lot on the north; thence S 01° 37′ 23″ E 169.00 feet along said lot line to the beginning of a curve concave to the northeast having a radius of 28.00 feet; thence southeasterly 43.98 feet along said curve through a central angle of 90° 00′ 00″ to the beginning of a compound curve concave to the north having a radius of 93.00 feet; thence easterly

50.69 feet along said curve to the beginning of a reverse curve concave to the southwest having a radius of 45.00 feet, a radial line through said beginning of reverse curve bears S 32° 51′ 13″ E; thence easterly and southeasterly 88.77 feet along said curve through a central angle of 113° 01′ 38″ to a radial line of said curve which bears N 80° 10′ 25 E; thence on the prolongation of said radial 44.43 feet to the southeast corner of said lot; thence N 00° 32′ 24″ W 175.00 feet along the east line of said lot; thence N 64° 22′ 24″ W 39.93 feet to a point of cusp with a curve concave to the northwest having a radius of 13.00 feet and to which point a radial line bears N 25° 37′ 36″ E; thence southeasterly, southwesterly and northwesterly 44.38 feet along said curve through a central angle of 195° 36′ 00″ to the beginning of a non-tangent curve concave to the north having a radius of 23.00 feet and to which beginning a radial line bears S 61° 39′ 09″ E; thence southwesterly and northwesterly 57.00 feet along said curve through a central angle of 142° 00′ 00″; thence on a non-tangent line N 74° 58′ 18″ W 101.96 feet to the point of beginning.

The above description gives you a running form of the different types and combinations of curves. Again, remember that it is most desirable to place the distance directly next to the direction, both in the case of a bearing for a straight line and the direction for a curved line. When you separate the distance from the direction, you open yourself to possible problems or questionable understanding, besides the need for additional words such as "a distance of," which would not otherwise be necessary. Again, this is a way to prevent verbosity and save paper space and typing.

Curve Controls

Another facet of curve control involves the juxtaposition of a curve which is non-tangent to its opposite boundary but tangent to the street line control. In Figure 38, such a configuration is shown for an area to be abandoned. The description can be written in either direction but each way must carry its particular ties in order to hold the correct relationship of the curve to the other lines.

Clockwise:

Those portions of Lots 152, 153 and 154 of the NEWMARK TRACT in the county of Orange, state of California as per map recorded in book 7, page 45 of Miscellaneous Maps in the office of the County Recorder of said county described as follows:

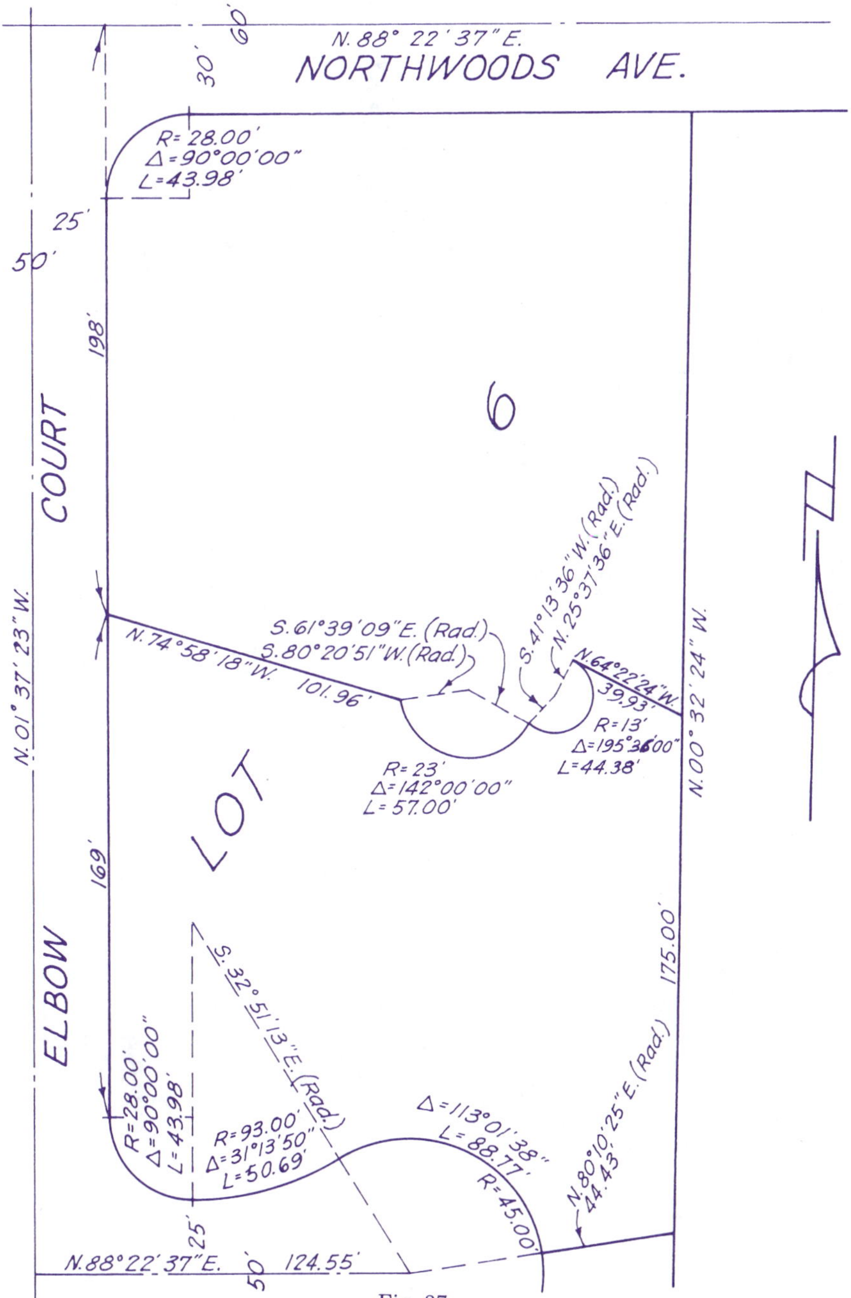
60'
30'
N. 88° 22' 37" E.
NORTHWOODS AVE.
R= 28.00'
Δ=90°00'00"
L=43.98'
25'
50'
198'
COURT
ELBOW
N. 01° 37' 23" W.
6
LOT
S. 61°39'09" E. (Rad.)
S. 80°20'51" W. (Rad.)
N. 74°58'18" W.
101.96'
S. 41°13'36" W. (Rad.)
N. 25°37'36" E. (Rad.)
N. 64°22'24" W.
39.93'
R=13'
Δ=195°36'00"
L=44.38'
R= 23'
Δ=142°00'00"
L= 57.00'
N. 00° 32' 24" W.
169'
175.00'
S. 32°51'13" E. (Rad.)
N. 80°10'25" E. (Rad.)
44.43'
Δ=113°01'38"
L=88.77'
R=45.00'
R=93.00'
Δ=31°13'50"
L=50.69'
R=28.00'
Δ=90°00'00"
L=43.98'
25'
N.88°22'37"E.
50'
124.55'

Fig. 37

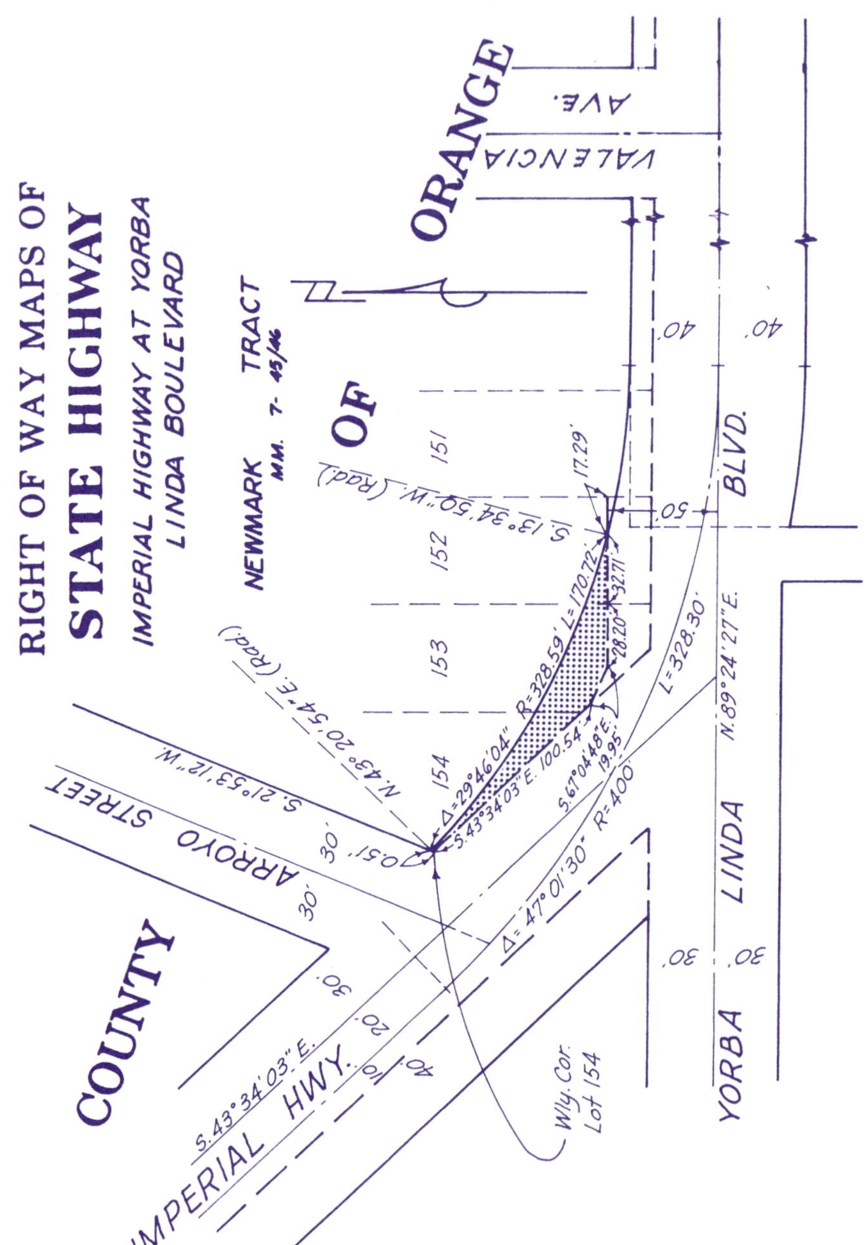

RIGHT OF WAY MAPS OF
STATE HIGHWAY
IMPERIAL HIGHWAY AT YORBA
LINDA BOULEVARD
NEWMARK TRACT
MM. 7- 45/46
COUNTY
OF
ORANGE
VALENCIA
AVE.
YORBA LINDA
BLVD.
IMPERIAL HWY.
ARROYO STREET
151
152
153
154
S. 13°34'50"W. (Rad.)
N. 43°20'54"E. (Rad.)
S. 21°53'12"W.
Δ=29°46'04" R=328.59' L=170.72'
S.43°34'03"E. 100.54'
S.67°04'48"E. 19.95'
Δ=47°01'30" R=400'
L=328.30'
N.89°24'27"E.
S.43°34'03"E.
0.51'
17.29'
32.71'
28.20'
Wly. Cor.
Lot 154

Fig. 38

Beginning at the most westerly corner of said Lot 154; thence along the westerly line of said Lot N 21° 53′ 12″ E 0.51 feet to its intersection with a curve concave to the northeast and having a radius of 328.59 feet and which is tangent to a line parallel with and 40 feet north of the center line of YORBA LINDA BLVD. as shown on said map and to which intersection a radial bears S 43° 20′ 54″ W; thence southeasterly 170.72 feet along said curve to a line parallel with and 50 feet north of said center line, a radial to said curve at said intersection bears S 13° 34′ 50″ W; thence along said parallel line S 89° 24′ 27″ W 60.91 feet; thence N 67° 04′ 48″ W 19.95 feet to the southwesterly line of said Lot 153; thence N 43° 34′ 03″ W 100.54 feet to the point of beginning.

Counterclockwise:

Those portions of Lots 152, 153 and 154 of the NEWMARK TRACT in the county of Orange, state of California as per map recorded in book 7, page 45 of Miscellaneous Maps in the office of the County Recorder of said county described as follows:

Beginning at the most westerly corner of said Lot 154; thence along the southwesterly line of said Lot S 43° 34′ 03″ E 100.54 feet; thence S 67° 04′ 48″ E. 19.95 feet to a line parallel with and 50 feet north of the center line of YORBA LINDA BOULEVARD as shown on said map; thence along said parallel line N 89° 24′ 27″ E 60.91 feet to a point on a non-tangent curve concave to the northeast and having a radius of 328.59 feet and which is tangent to a line parallel with and 40 feet north of said center line and to which point a radial bears S 13° 34′ 50″ W; thence northwesterly 170.72 feet along said curve to the westerly line of said Lot 154; thence S 21° 53′ 12″ W 0.51 feet to the point of beginning.

Note that the curve is held tangent to a line parallel with and 40 feet north of the center line of YORBA LINDA BOULEVARD in both cases.

Railroad Curves

A special type of curve to consider is that developed by the railroads based on their field method of establishing the center line for laying tracks. It consists of setting points along the curve at certain *chord* intervals (normally 100 feet) and using a constant deflection angle. This is why such a curve is designated as "a 1° curve," "a 4° 30′ curve," etc.; it means that "1°" is the deflection angle used at every 100-foot chord station and "4° 30′" is the deflection angle used at every station, etc. The radius is obtained from trigonometrical derivation. The commonly used chord definition is, a 1° curve has a chord of 100 feet subtending a central angle of 1°, while a 1° curve of arc definition (such as is used by some highway departments) has an arc of 100 feet subtending a central angle of 1°. In practice, 100 foot chords are used with deflection angles up to 7 degrees; 50 foot chords are used from 7 to 14 degrees; 25 foot chords are used from 14 to 28 degrees, and 10 foot chords are used for the larger degrees in sharp curves.* The reader is also referred to other standard texts on route surveying for additional information.

An easy way to visualize and remember the relationship between these factors is that a small angle has a long radius and a large angle has a short radius. With this in mind, you can understand how a spiral curve composed of a series of compound single curves with increasing, or decreasing, radii can be used for making an easy transition from a straight track into a curve section and then out to the tangent. Because high speed trains proved that the transition from a tangent line into the old original single simple curve and out to the tangent was too abrupt, the railroads developed and refined special curves in various combinations to be inserted for the transition. These are sometimes referred to as *taper curves* or *transition curves*.

In your use of these curves, remember that practically every railroad company through its engineering department has developed its own formula; therefore, be sure to check on the particular design in use by the railroad with which you are dealing.

The following copy of an example of transition curves is in conformity with tables prepared by the Atchison, Topeka and Santa Fe Railroad Company, and known in particular as "TAPER NO. 2." Only the beginning, middle and end parts are given because of the longevity of the entire curve.

Along a tangent curve to the right having a radius of 3,869.73 feet, through a central angle of 0° 27′ 00″ an arc distance of 30.39 feet;
Along a tangent curve to the right having a radius of 5,799.58 feet, through a central angle of 0° 18′ 00″ an arc distance of 30.26 feet;
Along a tangent curve to the right having a radius of 11,509.16 feet, through a central angle of 0° 09′ 00″ an arc distance of 30.13 feet;

*FIELD MANUAL FOR RAILROAD ENGINEERS, Nagle.

Tangent to said curve North 58° 35′ 35″ West 360.37 feet;
Along a tangent curve to the left having a radius of 11,409.16 feet, through a central angle of 00° 09′ 00″ an arc distance of 29.87 feet;
Along a tangent curve to the left having a radius of 5679.58 feet, through a central angle of 00° 18′ 00″ an arc distance of 29.74 feet;
Along a tangent curve to the left having a radius of 3,769.73 feet, through a central angel of 00° 27′ 00″ an arc distance of 29.61 feet;

* * * * *

Along a tangent curve to the left having a radius of 1223.27 feet, through a central angle of 01° 21° 00″ an arc distance of 28.82 feet;
Along a tangent curve to the left having a radius of 1,095.95 feet, through a central angle of 01° 30′ 00″ an arc distance of 28.69 feet;
Along a tangent curve to the left having a radius of 991.78 feet, through a central angle of 01° 39′ 00″ an arc distance of 28.56 feet;
Along a tangent curve to the left having a radius of 903.94 feet, through a central angle of 22° 19′ 57″ an arc distance of 352.33 feet; (this being the main simple curve and probably part of the original curve)
Along a tangent curve to the left having a radius of 991.78 feet, through a central angle of 01° 39′ 00″ an arc distance of 28.56 feet;
Along a tangent curve to the left having a radius of 1095.95 feet, through a central angle of 01° 30′ 00″ an arc distance of 28.69 feet;
Along a tangent curve to the left having a radius of 1,223.27 feet, through a central angle of 01° 21′ 00″ an arc distance of 28.82 feet;

* * * * *

Along a tangent curve to the left having a radius of 3769.73 feet, through a central angle of 00° 27′ 00″ an arc distance of 29.61 feet;
Along a tangent curve to the left having a radius of 5679.58 feet, through a central angle of 00° 18′ 00″ an arc distance of 29.74 feet;
Along a tangent curve to the left having a radius of 11,409.16 feet, through a central angle of 00° 09′ 00″ an arc distance of 29.87 feet;
South 79° 16′ 28″ West 264.47 feet;
Along a tangent curve to the right having a radius of 22,968.31 feet, through a central angle of 00° 04′ 30″ an arc distance of 30.07 feet;
Along a tangent curve to the right having a radius of 11,509.16 feet, through a central angle of 00° 09′ 00″ an arc distance of 30.13 feet;
Along a tangent curve to the right having a radius of 7,689.44 feet, through a central angle of 00° 13′ 30″ an arc distance of 30.20 feet;

(Margin annotations: This is Reverse of this — This is Reverse of this)

Notice the use of "right" and "left" because that is customary to railroads. Also see how it is necessary in this form to insert "an arc distance of" between the numerical distance and the numerical angle, thus adding three extra words to every course. A better form would be (for the first part):

> Thence northwesterly 29.87 feet along a tangent curve to the left having a radius of 11,409.16 feet through a central angle of 00° 09′ 11″;

and (for the last part):

> Thence westerly 30.07 feet along a tangent curve to the right having a radius of 22,968.31 feet through a central angle of 00° 04′ 30″;

As previously mentioned, the railroad type designation for curves by degrees is sometimes used in highway work with the exception, however, that the linear control is a 100 foot *arc* length instead of the 100 foot chord. If you are confronted with this situation, remember that your figures will be different when based on the arc.

PROBLEMS

1. Is it desirable to use four or more elements of a curve in a legal description?
2. Is the center of a circle the same as a mid-ordinate of that circle?
3. Explain the method of determining the direction of concavity of a curve.
4. In reconstructing a curve, what is the most acceptable control to hold?
5. Which "material items" will prevail in a description?
6. Name four sources of establishing a base of bearing.
7. There are seven calls generally used in the application of the "rules of construction;" list them in their order of superiority.
8. Is parol or extrinsic evidence admissible if it adds to or changes the meaning?
9. What are the two bases of azimuths in the U.S.A.? Give the control direction for each one.
10. Explain the method of determining the position and the relationship of the interior lines of "the southeast forty feet of the northwest sixty feet of Lot 28 . . . ." (Lot 28 is 100 feet square with the northwestly line bearing N 50° E)
11. Complete the following description for dividing Lot 5 shown in the drawing. Begin on "M Street" and describe the curves going north so that the last line is tangent to the last curve and ends at the point on the north line of the lot at 220 feet westerly from the northeast corner.

That portion of Lot 5 etc., etc., lying easterly of the following described line:

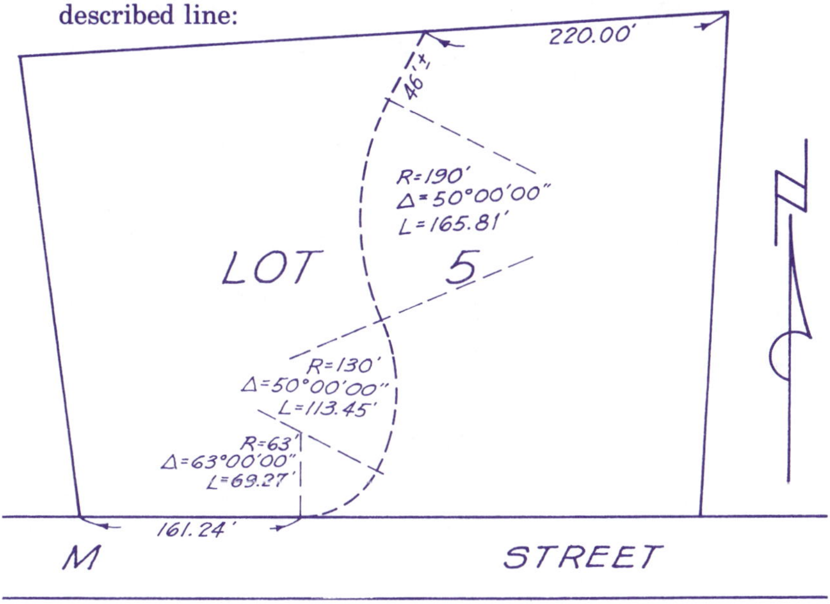

12. Plat the following descriptions in the LOT 7 shown here.

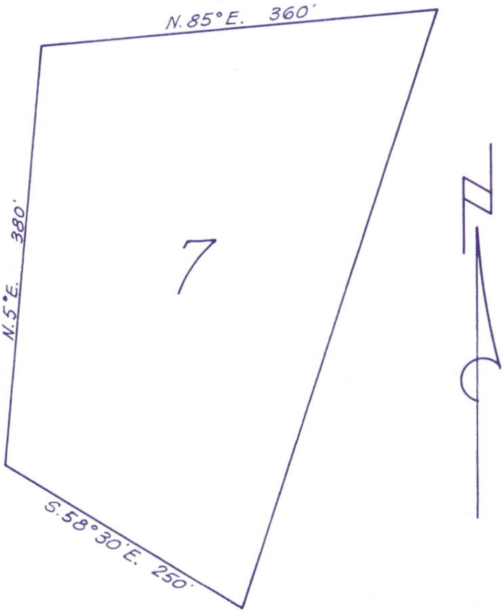

NOTE: Enlarging the drawing to a scale of 1″=100′ will make it easier to draft the parts.

a-The E'ly 120′ of the S'ly 80′ of Lot 7.
b-The S'ly 100′ of the N'ly 250′ of Lot 7. EXCEPT the W'ly 200′. ALSO EXCEPT any portion lying west of the easterly 100′.
c-The W'ly 110′ of the N'ly 150′ of Lot 7.
d-The W'ly 80′ of the S'ly 80′ of Lot 7.
e-The E'ly 200′ of Lot 7. EXCEPT the S'ly 80′. ALSO EXCEPT the N'ly 250′.
f-The S'ly 100′ of the N'ly 250′ of Lot 7. EXCEPT the E'ly 100′.
g-The S'ly 80′ of Lot 7. EXCEPT the E'ly 120′. ALSO EXCEPT the W'ly 80′.
h-The E'ly 100′ of the N'ly 150′ of Lot 7. EXCEPT the N'ly 25′.

13.

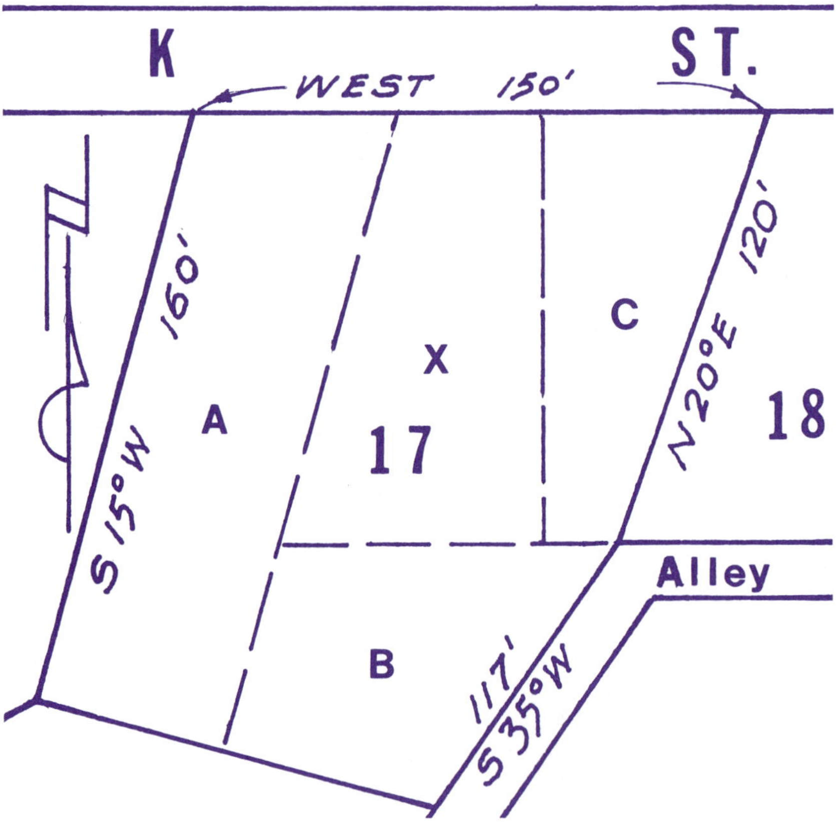

Lot 17 as per map is shown above. The approximate positions of the several descriptions given below are sketched thereon.

The following described parcels were deeded by the owner in chronological order as follows:

(C) Portion of Lot 17 described as beginning at the northeast corner; thence west on the north line, 60 ft.; thence south 112 ft. to the west prolongation of the south line of Lot 18; thence east to the southwest corner of Lot 18; thence N 20° E 120 ft. to the point of beginning.

(B) That portion of Lot 17 lying south of the west prolongation of the south line of Lot 18; except the westerly 50 ft. of said Lot 17.

(A) The westerly 50 ft. of Lot 17.

(X) Portion of Lot 17 described as follows:
Beginning in the north line distant west 60 ft. from the northeast corner; thence west 40 ft.; thence S 15° W 119 ft.; thence east 70 ft.; thence north 115 ft., to the point of beginning.

1. Do the descriptions cover all of Lot 17 without overlaps or gaps between the parcels? Explain fully by sketch and words.
2. The original owner is unavailable for any new deeds; therefore, if you conclude that any of the descriptions are incorrect or insufficient, determine a simple way to solve the problem with minimal documents and write a description which will adjust the lines, and name the proper person(s) to execute the adjustment deed(s).

ASSIGNMENT

Select several groups of curves from different maps, such as recorded Tracts, State Highways etc., and write your own descriptions following along the format given on pages 4.33 & 4.34. This will give you additional reference material and help you have confidence in writing curve descriptions easily.

— NOTES —

Chapter 5

Boundaries - General

Introduction

There are any number of descriptions of the subject apropos to the circumstances involved. Physical things are used where they exist or are to be placed; this includes either natural or artificial objects. The abstract form of a mathematical line without breadth is used between objects either previously or subsequently established for limits to that line.

"A boundary, in its stricter sense, is a visible mark indicating a dividing line between two things or it is that line itself; it marks off a given thing from other things like in kind as one field or country from another. A bound, on the other hand, is the limit or furthest point of extension of one given thing, that which limits it not being specifically considered." (Burrill L.D.)

Boundaries are also: "Every separation, natural or artificial, which marks the confines or lines of division of two contiguous estates. Civil Code, La., Art. 826. Limits or marks of enclosure if possession be without title, or the boundaries or limits stated in title deed if possession be under a title. Snelling v. Adair, 196 La. 624, 199 So. 782, 787." (Black L.D.)

There are several designations for "natural" boundaries. Bouvier's Law Dictionary states that, "a natural boundary is a natural object remaining where it was placed by nature." Other authorities define various natural boundaries thus: A savanna, or natural open meadow, "is a natural boundary in the sense in which that term is used in the construction of deeds." Stapleford v. Brinson, 24 N.C. 219 "A tract of land may itself become a natural boundary." Raymond v. Nash, 57 Conn. 447. If well known, a lot designated by number in a city is a natural boundary. Sayers v. City of Lyons, 10 Iowa 249. A meander line of a mountain ridge is even so considered, Bluffs of rocks and summits of mountains may constitute natural boundaries. Eureka Mining etc. . . . Co. v. Way, 11 Nev. 171. And, of course, streams, rivers, lakes, ponds, shores, beaches and trees are included among the works of nature if found on the land as they were placed by nature and are objects permanent in character.[10]

Artificial boundaries, sometimes called monuments, are considered in one sense as those established by man and yet such items as sidewalks, where the back edge is placed on the line between the street and the property, party walls, fences built of enduring material, highways and streets,[11] have been and are classed as natural because of their stability and permanence.

In making boundary locations, calls for natural or artificial objects will only prevail when they can be identified with reasonable certainty.[12]

Because there is, by the very means of determination of boundaries, interrelationship with and interdependence upon monuments, the latter will be treated in detail but more specifically in Chapter 8 rather than to intermingle them here.

In summation, a boundary is a line or markings, legal, natural or artificial, between enclosures if possession is without title such as a fenced area not in conformity with lines of title, or, between ownerships if title is recited in a document.

The describing of the boundary, in words which delineate, quantify and qualify the conditions by which the limits are to be established and held, is known as a land description. When words, or the lack of them, or the arrangement of them, used in descriptions result in ambiguity, both the interpretation of them and the physical location of the boundary lines from them become questionable.

Rules of Construction

It has been held that if the wording in a legal description is adequate enough for a competent surveyor to follow and lay it on the ground, with or without the aid of extrinsic evidence, then it is sufficient.[13]

Obversely, when a description is created from a survey, if any disagreement, inconsistency or element of contention is found in the field notes or between them and the calls in the deeds, then the rules of construction for ascertaining boundaries are to be invoked.[14]

[10]Miles Land Co. v. Hudson Coal Co., 246 Pa. 11, 13 & 23, 91 Atl. 1061, approving 4 Am. & Eng. Enc of Law (2d. Ed.) 746.

[11]Medara v. DuBois, 187 Pa. 431, 41 Atl. 322; Perich v. Maurer, 29 Cal. App. 293, 155 Pac. 471; Hoffman v. Port Huron, 102 Mich. 417, 60 N.W. 831.

[12]Carter & Bro. v. Collins, Tex. Civ. App., 192, S.W. 316, 321.

[13]Blume v. MacGregor (1944), 64 CA 2d 244, 251, 148 P 2d 656, 661.

[14]Gray v. King (1950), (Tex. Civ. App.), 227 S.W. 2d 872.

The intent of the parties to document, if the intent can be gathered from the language used, is the controlling factor even over the usual guides.[15] In Walsh v. Hill,[16] Judge Sanderson said, "The only rule of much value — one which is frequently shadowed forth, but seldom, if ever, expressed in books — is to place ourselves as nearly as possible in the seats which were occupied by the parties at the time the instrument was executed; then, taking it by the four corners, read it." Furthermore, this means, "It is the intention effectually expressed, not merely surmised."[17] A deed is to be treated in the same light as a contract giving full import to the intent if it can be ascertained.

In some situations, the grantor is favored and in others, the grantee is favored. The rule is that a private grant must be interpreted in favor of the grantee but that any reservation is held in favor of the grantor. On the other hand, every grant by a public body, or officer thereof, must be interpreted in favor of the grantor probably because the public body is always bound to protect the public interest for sake of future citizens.

In every case, consideration must be given to the circumstances existing at the time and place of the creation of the description. Some of these circumstances are extrinsic or outside of, i.e., not expressed in, the writing, hence they are "parol" or words not included in the writing in question.

The construing of the meaning of words and phrases used in a legal description will be made most heavily against the scrivener, especially if he is proficient and cognizant of such matters.

In some cases, it is permissible to introduce parol or extrinsic evidence consistent with matters in the writing to *assist* in construing the description. An example is where a call was made to a "junction" of a railroad and extrinsic evidence was admitted to determine which one of two junctions was intended.[18] Parol evidence is not limited to help in dual interpretation or obvious conflict; it can be brought into play for pure explanation to support the intention.

Parol or extrinsic evidence may not be admitted "to add to or vary a description that is definite and certain."[19] Nor may it be used to explain an apparent ambiguity in the conveyance.

[15]County of Los Angeles v. Hannon, 159 C 37, 112 P 878, Losee v. Jones, 120 Utah 385, 235 P 2d 132, 137, Carlisle v. Graves (La. App.), 64 S 2d 456, 461.
[16]38 Cal. 481, 487.
[17]Perry v. Buswell, 113 Maine 399, 94 Atl. 483, 484.
[18]County of Los Angeles v. Hannon, 159 Cal. 37, 112 P 878.
[19]Joerger v. P G & E, 207 Cal. 8, 32, 276 Pac. 1017, 1028.

So in writing descriptions, use words and phrases that clearly convey the intent of the parties concerned without creating any need to resort to extrinsic evidence for clarification.

Monuments as Boundaries

"A monument is some tangible landmark established to indicate a boundary."[20] Natural monuments refer to so-called permanent objects in nature, such as rivers, streams, ponds, lakes, beaches and shores.[21] Although trees, hedges, springs, creek deltas and rocks are of nature, they possess the habitude of being able to lose their permanency.[22] They become "lost boundaries" because the original location cannot be found.

Artificial monuments are things made by man or marks imposed upon things by man.[23] In the Pennsylvania Canal Co. case, an exact location was determined by a sharply defined cut in a rock bank on the side of the canal and the mark was taken as a monument.

Whether they are 1″ x 2″ stakes or 6″ x 6″ posts or iron (either rods or pipe) cut into certain lengths or posts made of concrete or granite cut into the shape of posts or "x" 's cut into the stone, etc., etc., they are classified as "artificial" in contrast with those considered above as "natural."

The phrase, "artificial monument," is sometimes also applied in a different sense in that it is not a true or correct representation of a corner. In one case where, even though a post existed between adjoining land owners, it was rejected under the circumstances of the case as an artificial monument because it was not referred to in the deed. Nor had any witness spoken of it as a monument.[24]

Even a house becomes a monument to the extent that it will include all the ground covered by the building and thereby fix the lines. This was held in the case where a house was referred to in the field notes.[25]

If a party wall is specifically called for in the description as a boundary, it is a monument according to the rule that monuments

[20]Whitcomb v. Milwaukie, 61 Or. 292, 296, 121 Pac 432.

[21]Hicks v. Coleman, 25 Cal 122, 85 Am D 103, Wheeler v. Spinola, 54 N.Y. 377, Goodson v. Fitzgerald, 40 Tex. Civ. A 619, 90 SW 898, Mansur v. Blake, 62 Me. 38, Dana v. Jackson St. Wharf Co., 31 Cal. 118, 89 Am D 164, Banks v. Ogden, 2 Wall 57, 17 L. ed 818, 69 US 57.

[22]Kidder v. Smith, (Mo. A.) 180 S.W. 1061, Perry v. Pratt, 31 Conn. 433.

[23]Pennsylvania Canal Co. v. Harris, 101 Pa. 80, Wise v. Burton, 73 Cal. 166, 14 Pac. 678.

[24]Law v. Hempstead, 10 Conn. 23.

[25]Wise v. Burton, 73 Cal. 166, 14 Pac. 678.

control calls for distances.[26] An unusual situation arose when a man sold the corner part of his property with the main portion of the description reading:

> Beginning at the southwest corner of said lot; thence easterly 75.00 feet along the southerly line thereof to the centerline of a brick wall to be built; thence northerly 100.00 feet along said centerline . . . . "

When the surveyor established the new lines, he set offset stakes 2.00 feet east of the 75.00 foot property line so that the mason would have room to build his wall without disturbing the stakes; however, the mason instead took the line of stakes as the property line and built the wall on them, making the centerline at 77.00 feet east of the southwest corner. When the seller discovered this mistake, he endeavored to get the buyer to pay for the additional two feet of land, but he encountered only stubborn rebuff. He therefore took recourse by filing suit against the buyer to recoup money for the extra land claimed by the buyer. The conclusion of the court was that since the description in the deed of conveyance stated, "to the centerline of a brick wall to be built," the physical location of the wall was to hold; case dismissed.

The lesson to learn here is that if a specific line position is to hold rather than a "wall to be built," then the control would be more effective by putting the wording about the wall in a parenthetical form such as,

> "thence easterly 75.00 feet along the southerly line thereof; thence at right angles northerly 100.00 feet, said line being proposed to become the centerline of a wall to be built . . . "

Should the surveyor find no survey monuments on a particular line in question but find a fence (by which is meant "any visible obstruction, whether a wall, a frame of wood and wire, or a hedge, interposed between two properties")[27] on a line position contrary to the line described in the deed, he would do well to ascertain under what conditions the "fence" exists, because the mere placement of a fence, even though permitted by the adjoiner, and without other facts to support, does not establish the fence as a boundary through any inference of an agreement.[28] The fence of itself is not sufficient to

[26]Medara v. DuBois, 187 Pa. 431, 41 A 322.
[27]Ogden's Calif. Real Property Law.
[28]Pedersen v. Reynolds, 31 CA 2d. 18.

establish a line of title[29] especially if it has not existed for too long a time.[30]

Agreement or Acquiescence

In looking at fence lines, disputed boundaries, etc., one must remember that there is a difference between "agreed boundaries" and "boundaries by acquiescence." An agreed boundary could be synonymized by classifying it as a compromised boundary; in other words, any dispute or doubt over the location of the line was settled by a compromise of the claims or positions of the concerned parties. On the other hand, acquiescence in and to a line is not based upon a dispute or a disagreement as to the line's location. In the case of Kirkegaard v. McLain[31], an arbitrary line on what both parties believed was the true line was agreed to for the purpose of building a fence and, without recourse to a survey to determine the true line, this might have constituted acquiescence if both parties were mistaken. But "acceptance" of a line not "known" was enough to support the trial court's conclusion that there had been an agreement to locate a "doubtful" boundary. In the appeal, the decision stated, "There is respectable written authority by Justice White, with hearing denied by the Supreme Court, holding in Rast v. Fischer[32] that where adjoining land owners, instead of having a survey according to the descriptions of their deeds, assume that stakes found by them denote the true line and accept it, their acquiescence is treated as a mistake and neither of them is estopped from claiming the true line; that evidence of an agreed boundary must disclose a valid preexisting agreement, and, to be valid, that agreement must have been based on a doubtful boundary line."

Referring again to "agreed" lines, the appellate court in the Kirkegaard case states: "In Clapp v. Churchill[33], it is stated that the doctrine of an agreed boundary line " . . . . rests fundamentally upon the fact that there is, or is believed by all parties to be, an uncertainty as to the location of the true line. When that uncertainty exists, or is believed by them to exist, they may, amongst themselves by agree-

[29]Wetherell v. Town of Newington, 54 Conn. 67, 5 A 858, Boltz v. Colsch, 134 Iowa 480, 109 N.W. 1106, West v. St. Louis, etc. R. Co., 59 Mo. 510, Sherman v. Brown, 160 App. Div. 859, 146 NYS 867.

[30]Cottrell v. Pickering, 32 Utah 62, 88 P 696, 699, 10 LRANS 404.

[31]199 CA 2d 484.

[32]107 Cal. App. 2d 129, 236 P 2d 393 (1951).

[33]164 Cal. 741, 130 P 1061.

ment, fix the boundary line, and that agreement will bind all the consenting parties." "***" "In Lusk v. Krejci (supra)[34], it was held that the doctrine of agreed boundary line did not apply where a wall was constructed by the plaintiff and the testimony showed that the parties never intended the wall to serve as a boundary, and the record was devoid of any evidence of a disagreement or uncertainty between coterminous owners as to the location of the boundary line."

In contrast, as long as adjoining owners hold their occupation on a line by agreement (either oral, written, express or implied), their condition is status quo and the excess is held by the same tenure as that holding the main body of land — but, if either party decides to transfer title to or from the agreed line, it will not hold without a document containing words of conveyance for that portion of the land bounded by the agreement which is not included within the description.[35] This is the type of condition which requires the showing of both lines by the surveyor, and the scrivener must await the decision of the adjoining owners as to what line is to be described (it might not even be either one of the surveyed lines) for inclusion in the document of agreement or for the transfer of the disputed area. Statutes of limitation vary from state to state, but it has been held that extended occupation of land beyond the true line of a deed does create a certain form of title ownership. However, where the boundary line of tenancy is not coincident with the line described in the deed of acquisition, the holder does not possess a marketable title until a new document is created to describe it and is properly executed between the parties concerned. Even in a case of quiet title when the court awards jurisdiction of a piece of land not specifically covered by a description in any previous document held by the awardee, it is necessary for constructive notice to have a new instrument to cover that area deposited in the office of public records, whether it be a new deed, a quitclaim or a copy of the court decree. When adjoining parties make a verbal agreement to accept the fence as their common boundary, there is no public notice, hence no marketability of title, although the possessory title is good as long as the adjoiners who made the agreement stay there. The courts have said that a fence of itself does not a line of title make . . . . a fence is made to keep animals and people in or out of a certain area.

Bringing this right home to our profession, no one has the legal

[34]187 Cal. App. 2d 553.
[35]Young v. Blakeman 153 C 477, 95 P 888 and Ogden's Calif. Real Property Law (1956).

right nor vested authority to change any line of a recorded title without a new and proper document in the recorder's office to support it.

There are times when adjoining owners find discrepancies themselves and in acknowledgement of the facts, record an agreement to correct them. For instance, the westerly line of a certain subdivision was not labeled on the record tract map and it had no ties to the westerly line of the section, but subsequent surveys of the section showed that this tract by monuments on the ground, supported by comparative measurements on the map, placed the tract's westerly line about eleven feet west of the center-of-section line. A review of the description in the deed whereby the subdivider acquired title showed that he was limited here by the west line of the southeast quarter of the section, and in similar confirmation, the party adjoining his land on the west was likewise limited. As stated, the adjoiners recorded a document which confirmed the west line of the said southeast quarter as their common line, notwithstanding the fact that the subdivision lot corners had been in their position *out of title* for over thirty years.

Another example of a misplaced boundary is shown on a Record of Survey Map which refers to a boundary line of a subdivision which has its southwest corner labeled on the Tract map as the southwest corner of the section with no monuments or ties shown to support it. Several years after the Tract map was recorded, a surveyor with more curiosity than the subdivider's man found the correct location of the section corner with ties to bearing trees still in existence, 124.2 feet west and 86.6 feet south from the position shown on the Tract map. This matter went to court and the later map was confirmed against the Tract map.

The following example is a combination problem. In establishing the boundary of a metes and bounds description of Mr. Meadowbrook's land, the surveyor found that the northerly line crossed an old fence line into Mr. Talltree's land. When Mr. Meadowbrook was so informed and he conferred with his neighbor, they decided to exchange deeds to make their title conform with the monumentation of the fence line. The surveyor was instructed to make the necessary measurements and prepare the descriptions for the parts to be traded. This was done and the respective documents were signed and recorded. Shortly thereafter, Mr. Talltree had the opportunity to sell his land, but, in the process of bringing the title to date, it was discovered that the portion he had acquired from Mr. Meadowbrook in the exchange had been, and still was, encumbered by a mortgage. So remember that any movement of lines should give full consideration to the possible existence of related encumbrances and how those encumbrances, if any, are to be handled.

This matter of multiple vested interests must be considered very carefully by the scrivener, both when interpreting descriptions already on record and when creating new ones. If a call is made to a canal, a railroad, a street, alley or highway, it is generally assumed to be to the center line of that strip. There are situations, however, where the ownership of that strip was created as a fee title[36] and as such is not susceptible to ownership by adjoiners, although it may be encumbered by easements, mortgages, etc. Because of this, the scrivener must make proper and sufficient search of title to determine the true conditions. It might even be possible that the description ties by distance to the center line of such a strip, but because of the fee title, it will be necessary to either shorten the descriptive tie by proper mathematics, or add an exception to the description to withhold the part in fee.

Descriptive Courses

Flat calls are just the citation of a direction and distance with no qualification and no tie to anything at the end of the course. This is sometimes called a mathematical line because of the absence of physical attributes. Otherwise, calls include an assignment of a condition to either the direction or the distance, or both, or to the terminus of the course. Examples of the latter are:

1. "thence along the southeast line of said lot, N 38° 24′ E 132.67 feet;"
   This requires following "the southeast line of said lot" regardless of the bearing but only for the amount of the distance cited.
2. "thence S 02° 14′ E 87.16 feet for the length of the building;"
   This requires following the bearing for whatever length the building proves to be.
3. "thence N 16° 32′ W 378 feet along the center line of an existing wall to the end thereof;"

[36]Terms "fee," "fee simple" and "fee simple absolute" are equivalent; Boon v. Boon, 348 Ill. 120, 180 NE 792, 794. Fee simple signifies a pure fee;. . . . .It is the largest estate and most extensive interest that can be enjoyed in land. Haynes v. Bourn, 42 Vt. 686 and others. Black's L.D.

This requires both the bearing and distance to give way to the actual physical position of the wall.

4. "thence N 41° 52′ E 192.16 feet to a point on the southwest line of Higgins Tract which is distant southeasterly thereon 51.78 feet from a granite post at the most westerly corner of said Tract;" This of course requires the line "N 41° 52′ E 192.16 feet" to give in entirety to the position of the point described on the southwest line of Higgins Tract.

In writing descriptions from the work of a surveyor, several questions must be considered. Are the surveyed lines free lines? That is, are they new creations in an open area non-dependent upon boundary conditions? Or, do they follow in whole or part existing lines of record ownership?

If you are dealing with "free lines," then you should follow right along the surveyor's footsteps on the assumption that he established those lines per instructions from his client. On the other hand, if any part of the area to be described is bounded by lines of record title, you should research them and be sure you have all such information (maps, deeds, etc.) to use for reference and ties in your new description. Based on this, if there are questions concerning the survey information given you, inquiry into the surveyor's field notes would be in order and would simulate a court's procedure if questions arose in a subsequent dispute. 12 Am. Jur. 2d 644, Boundaries §111 states, "The field notes and plats of the original surveyor are primary evidence as to the true location of boundaries."[37]

However, if the surveyor perchance did not recognize and therefore did not adhere to a boundary pertinent to his survey, then the scrivener with his background information should take due note and make proper ties to the boundary in his description or except out of the description any part of such boundary land.

For safety and security in writing descriptions, boundaries of whatever sort, natural, artificial or legal, abstract or concrete, should be incorporated as much as possible as an integral part of each step around the perimeter of the subject area.

Returning to lines again, if the identification of a line is only its direction and distance with no controls attached along the way nor to the end of it, it is called a "free" line. Such lines may be created either by a surveyor in the field or by a draftsman or scrivener in the office.

[37]Citing Taylor v. Fomby, 116 Ala. 621, 22 So. 910, Alexander v. Lively, 5TB Mon. (Ky.) 159, Ayers v. Watson, 137 U.S. 584, 34 L ed. 803, 11 S. Ct. 201.

Upon recordation, these subsequently become "fixed" and as such carry the status of "boundary" lines.

The identification of lines develops from characteristics that are either inherent or applied. In other words, if the boundary line follows a river, the *inherent* character of the qualifications assigned to a river applies to the interpretation of that boundary; if the qualification is assigned to the line that it be and that it follow, "the southeast side of Jones' building and the prolongation thereof," then that *applied* characteristic controls the line.

Consider for the moment the analysis to be applied to the "relocation" of a line vs. the "reestablishment" of a line. Webster describes "relocation" as "a second location." Black refers it to mining law which considers it as "A new or fresh location. . . . .to change the boundaries or to correct mistakes in the original location." Webster's definition of "reestablish" is, "to establish again in or to a former place, position or state; set up, fix or confirm again."

The essence of this is that if an analysis of a line reveals that it is a *relocation* of one already on record, you cannot accept it per se unless or until a description of the new relocated line is properly recorded. Conversely, a *reestablished* line is one correctly replaced into the position where it was created and recorded. Be certain of which kind of a line you have before you incorporate it into your description.

Bearings

A bearing is the direction of a line stated specifically or generally; i.e.: N 39° 31′ 46″ E, or northeasterly, and it is usually in reference to a meridian or a base line. Bearings are classed as *true, magnetic* or *compass,* of which the last two are in some respects synonymous. In the early days of the colonial descriptions, it was common practice to use bearings based on compass surveys. When the continuous annual change of the local magnetic declination is considered and you must analyze the development of a description having bearings derived from compass readings, the date interval times the rate of change must be applied in order to convert present day compass readings to the conditions existent at the time the description was created. Additional caution should be applied to such early descriptions because of local variations and the unreliability of the then reported declinations. The early government surveys of public lands were also based on the compass, which was deflected in some areas as much as 20° by local mineral conditions. In 1902, the general land office issued rules that all subsequent government surveys would be based on directions determined by reference to a true meridian without reliance on a magnetic needle. This was made possible by an improved instrument called the solar compass.

In using compass bearing lines, remember that the variations from true north of the lines of magnetic force are measured usually by the declination of the north-pointing end of the compass needle east or west of true north and this variation must be applied throughout in order to achieve the correct relationship of the boundary lines, especially where you have one area based on compass readings and an adjoining area based on a true meridian.

Whereas bearings derived from the compass are usually written from the successive readings of the needle, bearings developed from a true meridian, or from a base of another true bearing, are the result of calculations made from angles turned between lines.

In the application of bearings to the study of boundaries of either an independent parcel or of adjoining properities, the keynote is "relationship." Regarding one single parcel, every bearing may be related to the one original bearing cited for control or there may be changes of bearing control introduced at different points along the traverse. In former times, descriptions were occasionally written with both the bearings for successive lines and the angle between them cited; in this way, one could compare the cited angle with the angle calculated from the bearings for a check-out. If a description is written by taking the bearings from a calculated and closed traverse sheet of the survey, there should be no question of error other than miscopy.

Turning to the consideration of bearings on lines of adjoining lands, you must be aware of the basis of bearings used in the adjoiner as compared with that for the instant property. Again it is "relationship." If you are fortunate enough to have a common basis, it is simple but this is the exception rather than the rule. Furthermore, you may not look at the bearing in just the first course of the deed for your neighboring parcel and accept it per se because there could be any number of alterations of that base between the beginning and the point of contact with your piece.

If the resultant direction of bearing development of a tied-to line in the adjoining parcel description bends the common line described in the instant parcel, you may have a problem with the succeeding lines. For example, Figure 39 shows Parcel A, courses 1 through 8, as a mathematically closed traverse, but that course No. 5 ties to a boundary which by its interior control forces that course southwesterly from its pivot, the common point 5, at the southerly terminus.

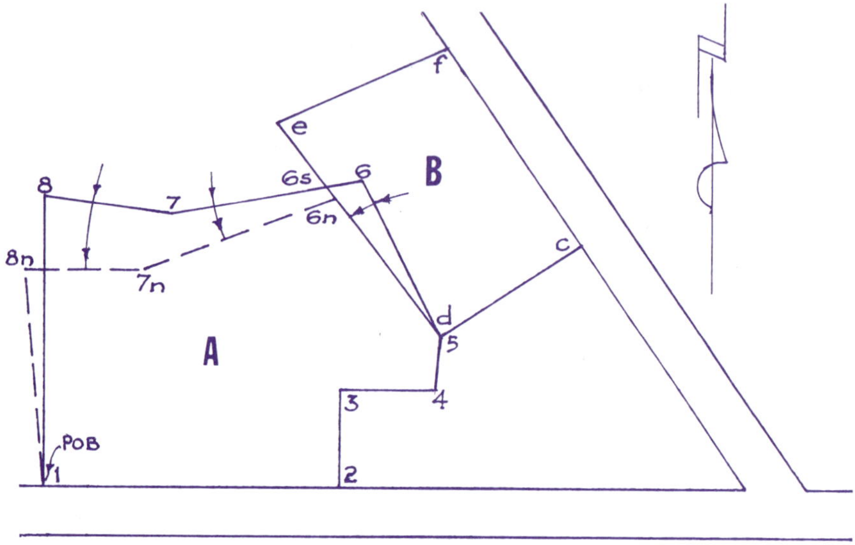

Fig. 39

Because this course is tied to course d-e in Parcel B, the basis of bearings are here altered and the subsequent line-angular relationship is held from this base shown as 6n, 7n and 8n which shorts Parcel A on the north and extends it on the west. This is one reason for studying boundary deeds *before* running a survey or writing a description.

Distances

In relationship to boundaries, distances are controlled in one of three ways:

1) By a line going to one which determines by its position the distance which will hold;
2) By a specified length on or along a boundary line which becomes also the length of the instant line, or;
3) By a point on a line at a specified relation to a control point.

The first and third way is a simple "stop" to the line and the resultant measurement on the ground is the answer for the distance. The second way depends upon the manner in which that particular boundary line is described. If it is a flat call or if it is tied to another tie, there is no problem. However, should the boundary line be dependent upon a questionable condition such as the application of the apportionment theory vs. the remnant theory, then further study of the adjoining situation must be made prior to a decision on the instant property. This last subject matter will be discussed further in Chapter 7 on Analysis of Descriptions.

Quantity as a Boundary

Recognition must be given first to the fact that quantity in a description may be used in one of three ways:

1) It can be positively definitive such as "the north fifty feet of Lot 2 in Block 6 of the Westside Tract";
2) It can be dependently definitive such as "the north 10 acres of the west half of the northwest quarter of the northeast quarter of Section 7" where the north-south distance for that 10 acres depends on the east-west distance developed from the field survey breakdown of the section; and
3) It is stated as a supplementary fact at the end with no finite control.

The first two ways do control the positioning of the acreage boundary. The third way does not generally control but is sometimes used as a check against an error in traversing or it may exert some influence in the determination of its boundary in absence of satisfactory calls.

As with other ties to boundary conditions, the call to a quantity control demands the physical establishment of the fact for locative purposes.

Essentially, when quantity is used as a boundary, the limiting line is a mathematical line; i.e., "bounded on the north by the south line of the north fifty feet of Lot 2. . . ."

Quantity can also be used as a minus factor expressed as an exception; i.e., "EXCEPT the north fifty feet of Lot 2. . . ."

When quantities are used as a boundary exception to the whole, another phase of title ownership becomes involved and particular attention must be given to the form used. Whether you are interpreting a recorded description or creating one to be recorded, you must recognize the wording format which differentiates between a fee excepted or a fee not excepted. The side effect especially pertinent to the subject is, when there is a *condition subsequent,* the question is whether the *rights* linked with that condition subsequent pass to the grantee or not; in other words, will the grantor or grantee benefit from the reversion of that parcel of land if it is not used for the purpose specified in the deed.

A general statement can be made to the effect that if the original grant was a defeasible fee or easement for a specific purpose, then later the entire original holding of land is conveyed with an exception of that area reciting the qualification which was attached to it, the fee in the excepted area also passed but subject to the conditions which had been imposed upon it. Although it is stated that the individual grantee is usually to be favored as against the grantor,[38] and in an unclear conveyance, an interpretation which would deprive the grantee of his intended use of the property will be rejected, thus "favoring" the grantee, it appears that in California, reservations and exceptions are generally construed as being for the benefit of the grantor, especially if ambiguous.[39] Where exception is made, excepting interests "as heretofore deeded," the conditions stated in that previous deed prevail.[40]

Obversely, when a large parcel was granted, "excepting the part occupied by the right of way of the Iowa Central Railroad Company," it was held that the grantor retained the fee in the area described in the exception, even though the strip had been granted as an easement.[41] Apparently reference to a specific area to be excepted, rather than a condition or status of title, has been considered more in the light of permitting the grantor to retain any reversionary rights.

The scrivener is again admonished to review the codes and court cases of his area concerning these hairline interpretations and even to seek the counsel of a good attorney or title company.

Maps and Plats as Boundaries

Control of lines by reference to maps[42] may be to any part or in toto. Again, chronological relativity in regard to any survey work done prior to or subsequent to the creation of the map, the use of the map for conveyancing purposes and occupation developed therefrom is to be carefully considered before deciding which conditions will control.

[38]Derby v. Hall (Mass. 1854), 2 Gray 236, 250, 68 Mass. 236.

[39]Seligman v. Carr (1908), 8 CA 572, 97 P 324. Also Cal. CC Sec. 1069.

[40]Roxana Petroleum Corp. v. Jarvis (Kan. 1929), 273 P 661, Barker v. Lashbrook (Kan. 1929), 279 P 12.

[41]Hall v. Wabash R.R. (Iowa 1907), 110 NW 1039.

[42]Use of the word "maps" herein will include reference to "plats" also.

Maps showing undeveloped streets, so-called "paper" streets are good guides for ascertaining the parties' intentions and, if harmonious with those intentions, will control a conveyance influenced by the street as it exists on the ground.[43] On the other hand, if a street is laid out on the ground and improvements, such as fences or curbings exist by which a reference to the physical street on the ground is made as a controlling call, the map will be disregarded.[44]

Reference to either the boundary of a map or an internal tie shown thereon for control of the instant property will be held to be on the basis of either the survey performed for the creation of the map, or, if monuments and survey data are non-existent, then on the basis of the map per se.

Mention of a map, whether of public record or not, but preferably of public record, carries the same status *as a monument.*[45] But, on the theory of accepting the lines and physical monuments set out by the surveyor, the bearing and distances shown on the map do not carry the same prestige.[46] However, with today's use of electronic measuring instruments and one-second theodolites in the field and sophisticated computers in the office, the discrepancies referred to in the last case are not so common; therefore, the use of a map as a *monument* is more plausible.

Particular Calls

Streets and Highways

There are innumerable court cases which cite and support the rule that the conveyance of land adjoining a street or highway carries the underlying fee title to the center or to the original dividing line wherever it may lie within the street or highway. A claim based on this rule is subject to two conditions: 1) the easement and rights of the public, and 2) if the grantor owned the part in the street or highway at the time of the conveyance and, if so, there were no words denying the conveyance thereof.

Where reference and tie is made to "a street as opened," the physical location of the street with its limits will apply in contrast with reference to "a street as shown on the map of . . . . ." where the facts given on the map will control.

[43]Reid v. Klein, 138 Ind. 484, 37 NE 967.
[44]Philadelphia etc. v. Hinkle, 64 Pa. Super. Ct. 495, Singer v. Mayor etc. of City of N.Y., N.Y. 47 App. Div. 42, 62 NYS 347.
[45]Wilson v. Chicago Lumber & Timber Co., 143 Fed. 705, 711.
[46]Smith v. Harbaugh, 216, Ala. 202, 112 So. 914.

A mere tie across the street, such as "the southerly line of said parcel being parallel with and 93 feet south of the north line of A Street," does not of itself convey title in the street.

The question of whether the conveyance carries any title to any part in the street is a matter of construction, the determination of which is based upon the inherent language and the circumstances attached thereto which must be considered in deriving the intent.[47] Again, it is a matter of not immediately accepting what appears before you but of researching the record and conditions affecting the conveyance. And, in writing descriptions that have any part concerned with streets or highways, research the facts and describe them correctly in ways that leave no ambiguity.

The use of certain combinations of words which either perfect or cut off the conveyance of fee title in the street is explained in Chapter 9 on Streets.

Railroads

Ties made to So and So's Railroad or to the XYZ Railroad Company's right of way require as much, or more, delving into the interpretation of the conveyances than do streets and highways. Although it is generally conceded that the status of railroad right of ways is not the same, there are some similarities. Going back to fundamentals, if the railroad property is fee, any adjoiner is limited to the sideline thereof, but if the right of way is in fact an easement, the grantees on both sides are vested with fee to the center.[48] This brings up the matter of discernment as to the wording in the right of way description.

There are generally two types of such descriptions. One follows the form, "a strip of land 100 feet wide lying 50 feet on each side of the following described line:" in which case the right of way conforms with that line regardless of the physical location of the track(s). The other form reads, "a strip of land 100 feet wide lying 50 feet on each side of the center line of the located tracks," or, in some cases, ". . . . the center line of the tracks to be laid." In this situation, the physical location controls and establishes the position of the right of way and any adjoiners need a survey to determine their physical measurements to, and limits of, that boundary; however, in later years where the railroad company has realigned the physical location of the

[47]Graham v. Stern, 168 N.Y. 517, 61 NE 891, Bissell v. New York Central R.R. Co., 23 N.Y. 61, Huff v. Hastings Express Co., 195 Ill. 257, 63 NE 105, etc.
[48]Center Bridge Co. v. Wheeler etc, Co., 86 Conn, 585, 86 A 11.

tracks to accommodate high-speed trains, particularly in changing the original simple curve to spiral or transition curves, one must remain cognizant of the fact that the location of the *original* tracks established the position of the right of way and the surveyor must reestablish that original layout.

Thus, the scrivener, understanding the form of creation, must adhere to that in making ties thereto or deriving boundaries therefrom.

Private Ways

The conveyance of parcels of land bounded by a private way is generally held to carry to the center line thereof,[49] even though the granting clause includes its use by the grantor, his heirs and assigns.[50] An opened private alley has the same status as a public road as a boundary with conveyances of fee going to the center.[51]

PROBLEMS

1. Does a simple "agreed" boundary confer on the adjoining lands the status of a "marketable" title?
2. When does a free line become a fixed line?
3. Is a "relocated" line more synonymous with a "reestablished" line or an "agreement" line?
4. Is the grantee favored by a public grant?
5. What are the two types of railroad deeds?
6. What gives "identification" to lines?
7. Is a wall as a boundary called for in a deed as a condition subsequent sufficient to hold over a distance cited?
8. Does a fence absolutely establish a boundary?
9. If there is ambiguity in the wording of a description, which factor will control? — Application of the rules of construction or intent?

[49]Gould v. Eastern R. Co., 142 Mass. 85, Pitney v. Huested, 8 App. Div. 105, 40 NYS 407, Pettibone v. Hamilton, 40 Wis. 402, Paine v. Consumers' Forwarding, etc., Co., 71 Fed. 626, 19 CCA 99.

[50]Motley v. Sargent, 119 Mass. 231, Stark v. Coffin, 105 Mass. 328.

[51]Saccone v. West End Trust Co., 224 Pa. 554, 73 A 971, Carter v. Lebzelter, 45 Pa. Super. 478.

— NOTES —

Chapter 6

Water Boundaries

Introduction

Riparian owners are those whose lands border upon a river, stream, lake or pond. Land lying beyond the natural watershed of a stream is not "riparian."[52] Also, if there is any strip of land between

[52]Bathgate v. Irvine, 126 Cal. 135, 58 P 442, Town of Gordonsville v. Zinn, 129 Va. 542, 106 S.E. 508, 513, 14 A.L.R. 318.

the owner's land and the legal boundary of the water, that owner has no riparian or littoral rights.

Littoral owners are those whose lands border upon the sea, ocean or Great Lakes. However, usage has broadened the concept of riparian to the extent that littoral lands and rights are often described as riparian.

In regard to artificially-created water boundaries, such as reservoirs, the boundary owner is a riparian owner only if, as and when he meets the tests of whether he has any riparian rights applicable to the reservoir by express terms in the grant to the owner.

The *rights* of riparian and littoral owners are those rights which are acquired by law as an incident of the ownership of riparian and littoral lands, or by conveyances in the chain of title, to the bed, bank, shore or water of the boundary waters. Such rights may include the right to wharf out to deep water, the use of the water for domestic or commercial purposes, the use of the surface of the water for boating, fishing, hunting and other recreational purposes, the right to maintain certain aesthetic values, etc.

We will not go into detail here concerning the derivation and application of all the rights involved in riparian or littoral ownership. For a thorough and comprehenvise presentation of the views held and practiced by the Federal authorities, the reader is directed to Shore and Sea Boundaries by A. L. Shalowitz, 2 Vols. Because much has been written concerning this subject matter which you yourself can research, the objective here will be to direct your attention to those aspects which particularly affect the analysis of descriptions on the one hand and the writing of them on the other, and those items attendant upon them, based on such alternate conditions as: tidal or non-tidal waters; navigable or non-navigable waters; accretion or avulsion; alluvion or reliction; mean low water or mean high tide, and if the latter, which mean high tide? Some of these are pointed out in Figure 40. The engineer's or surveyor's primary concern is ordinarily only the boundary of the riparian or littoral owner's land to be surveyed or described which, by law or conveyancing, may be the ordinary high water, the ordinary low water, the middle of the lake, the thread of the stream or some other line identified in the chain of title.

Riparian or littoral laws vary from state to state. Since the ownership of the beds of most inland navigable waters were acquired upon statehood as an incident of sovereignty, whether such waters were navigable or not, for title purposes upon entering statehood, is a matter of federal and not state law. Thus, each of the aspects of riparian and littoral ownership in the state in which the property is located must be researched in detail and no matter which type of

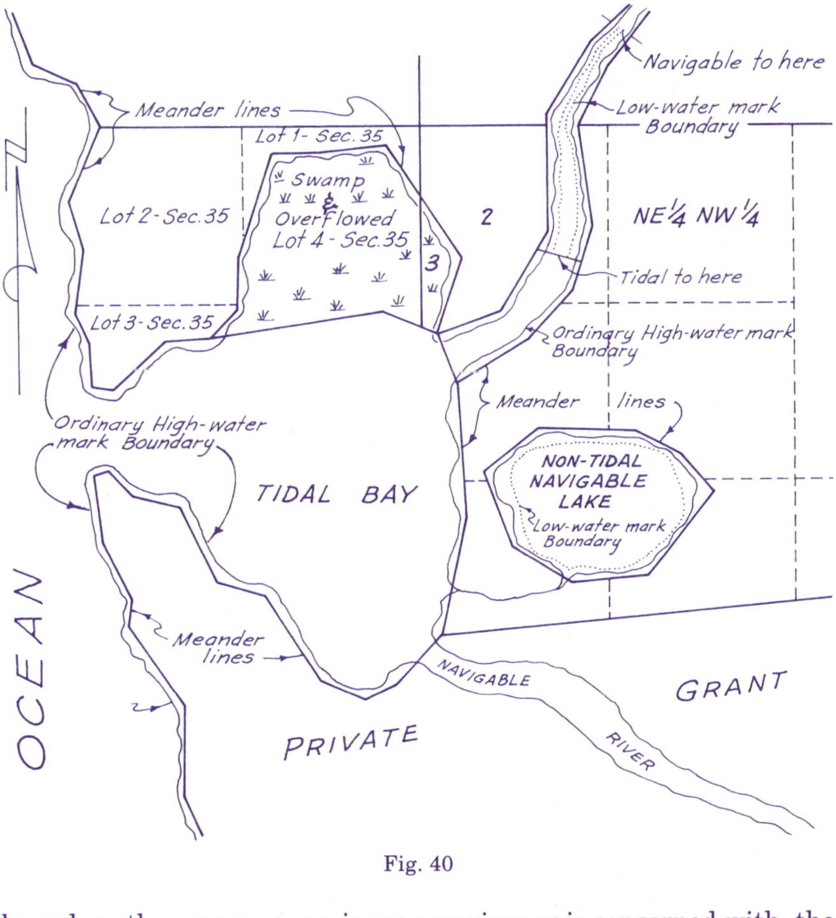

Fig. 40

boundary the surveyor, engineer or scrivener is concerned with, the rules must be applied to determine the extent of ownership so that in locating those boundaries and writing descriptions, nothing is taken away or added to that which the client truly owns. If you cannot determine such ownership, call upon the help and advice of an attorney who specializes in such matters or a title company that will tell you how much of the land they would insure.

Discussion of some of the principles of such laws and the problems in connection therewith will include the following:

Tideland boundaries,
Head of tidewater,
Ordinary high water mark — tidewater,
Meander lines,
Submerged lands,
Ordinary high water mark — fresh water,
Ordinary low water mark — tidewater,
Ordinary low water mark — fresh water,
Center or thread of fresh water lake or stream,
Projected boundaries into fresh water lake or stream,
Projected boundaries into tidewater,
Accretion, erosion and reliction,
Avulsion,
Inclusion of accretion in upland conveyances.

Tidelands

Tide is the ebb and flow of the sea[53] and tidal in reference to a river is not necessarily where the water should be salt but the spot where the tide, in the ordinary and regular course of things, flows and reflows. (Black's L.D.)

Tidelands, or tidal boundaries, by definition, are those parts of the shore or beach which lie between mean (or ordinary)[54] high tide and mean (or ordinary) low water. The limits of these two conditions are determined by both vertical and horizontal considerations as shown in Figures 41a and 41b. Figure 41b illustrates how the horizontal difference is proportionately greater in areas of lesser slope in contrast with Figure 41a.

Title to the tidelands and the submerged soil adjoining it to the

[53]Baird v. Campbell, 67 App. Div. 104, 73 N.Y.S. 617.

[54]"The word 'ordinary' when applied to a high or low water mark has generally been used in the sense of average by the courts of this country and of England." A. L. Shalowitz, SHORE AND SEA BOUNDARIES. Also, "ordinary" in connection with tides is equivalent to "mean"; Schureman, TIDE AND CURRENT GLOSSARY 23.

[55]Except that Texas and Florida by respective cases were confirmed to a limit of three marine leagues in the Gulf of Mexico. In U.S. v. Maine, Mar. 17, 1975, the U.S. Supreme Court found that the United States, rather than the Atlantic coastal states, has sovereign rights over the seabed and subsoil of the Continental Shelf of the Atlantic Ocean and based on the Submerged Lands Act (see Note 56 infra), "the United States has paramount rights to the seabed beyond the three-mile limit." Thus, Florida has a different limit on the east.

seaward limit of three geographical miles[55] is in the various states and individual owners under them.[56] It is subject, however, "to the servitude in respect of navigation created in favor of the Federal Government by the Constitution."[57] The public rights of commerce, navigation and fishery are upheld and delineated in many decisions.[58] The ownership of these lands is, in effect, a trust subject to, and for the benefit of, the rights of the public. In general, no Acts authorize the sale of submerged lands.[59]

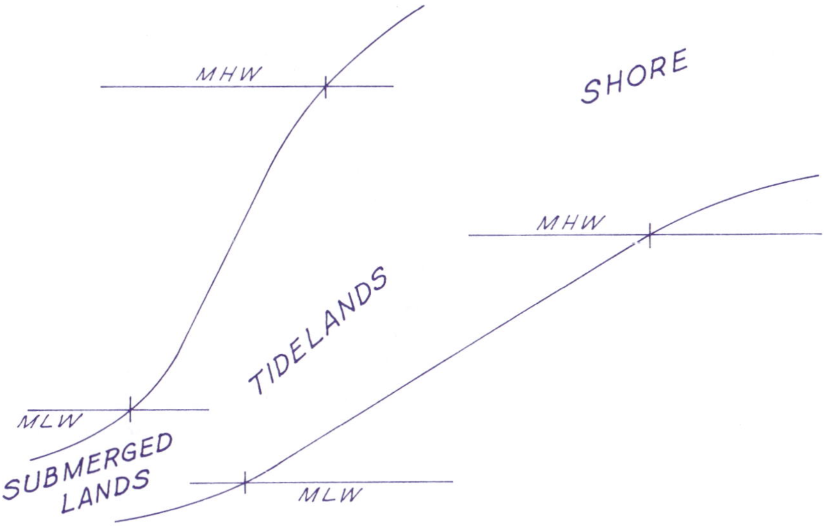

Fig. 41a Fig. 41b

The vertical difference according to federal law is developed from averages of all tides over an 18.6-year cycle, which is the period of the moon's complete oscillation in relationship to the sun and earth. The juxtaposition of the three objects affects tidal conditions in different ways at different places. For example, the high and low water differ-

[56]According to rules set forth in the Submerged Lands Act (Chapt. 65 Public Law 31) which was H.R. 4198 of the 83rd Congress, First Session, signed into law May 22, 1953 and found in 1953 U.S. Code and Congressional Service Vol. 1, p. 28-34.
[57]Paramount rights of U.S. defined in Gibson v. U.S., 166 U.S. 269, 41 L. Ed. 996.
[58]People v. Cal. Fish Co., 166 Cal. 576, 584, Illinois C.R.R. v. Illinois, 146 U.S. 387, 452, 36 L. Ed. 1018, 1042, Newcomb v. City of Newport Beach, 7 Cal. 2d. 393, 401, etc.
[59]A special Act of the California State Legislature did give certain tide and submerged lands to San Francisco for its waterfront development.

ential is as much as 34 feet in Anchorage, Alaska, a little under 6 feet at Los Angeles, 27 feet in the upper reaches of the Gulf of California, only 2 feet at Pensacola, Florida, and 7 feet around New York City.

In some cases, a 19-year cycle is considered a practical approach because the fourth effect on tides, meteorological conditions, are non-periodic. Although this assumption is acceptable for isolated citations or closely time-related comparisons, a 76-year-apart comparison (4 cycles) would be 1.6 of a year, or 19.2 months, off from four 18.6 year cycles.

In California, certain court rulings[60] have held that the mean high water mark should be the mean of the "neap" tides, but in Borax Consolidated, Ltd. v. Los Angeles,[61] extensive reference was made to the Coast Survey Special Publication No. 135, Tidal Datum Planes, wherein it states that "Mean high water at any place is the average height of all the high waters at that place over a considerable period of time," and that "an average of 18.6 years should be determined as near as possible." There is another objection to the use of the neap tide rule which prevents it from being acceptable on a universal basis: the principal variations in the rise and fall of the tide are related to three conditions which even vary in different regions: (1) the moon's phase; (2) parallax (distance of moon from earth); and (3) the moon's declination (distance north or south of Equator).[62] In United States v. Washington (State)[63] the tests in the Borax case for determining the limits of a grant by the United States were used and upheld by the Supreme Court's refusal to review the Circuit Court's ruling.

The general conclusion to be drawn from the foregoing is that in most states, the ocean boundary of title of federal lands is the line of mean high water[64] established on the basis of an 18.6-year average. For example, Lot 2, Section 35 (Figure 40).

Meander Lines

Survey lines used to delineate water boundary areas are called meander lines. Presumably, these approximate the water line, but

[60]Teschemacher v. Thompson, 18 Cal. 11, 79 Am. Dec. 151, and Forgeus v. Santa Cruz County, et al., 140 Pac. 1092.

[61]296 U.S. 10, 22 (1935), explaining in footnotes the meaning of "neap" as "nipped" and "spring" as "fuller" tides.

[62]Marmer, TIDAL DATUM PLANES 5, Special Publication No. 135, USC & GS (1951).

[63]294 F. 2d 830 (1961).

[64]This has been modified in a few states where upland ownership is carried to the low water mark, Shalowitz, SHORE AND SEA BOUNDARIES.

variations of all amounts are found in comparing the physical conditions with the surveyed lines; however, this does not alter the line of ownership which follows the edge of the water as previously discussed. Meander lines are not boundaries defining the area of ownership next to water but are run to define the sinuosities of water lines and for the purpose of ascertaining the quantity of land embraced within the survey for patent.[65] This applies not only to ocean front boundaries but to bay shore lines, lake perimeters and the sidelines of rivers.

California

The boundaries of the tidelands of California under jurisdiction of the State as her sovereign right is another story. Forgeus v. County of Santa Cruz cites the definition given in Teschemacher v. Thompson[66] as the prevailing rule: "By 'ordinary high water mark' is meant the limit reached by the neap tides, that is, those tides which happen between the full and change of the moon twice in every twenty-four hours." Figure 42 shows the juxtaposition relationship between the federal high and low water lines and the neap tide lines. There has

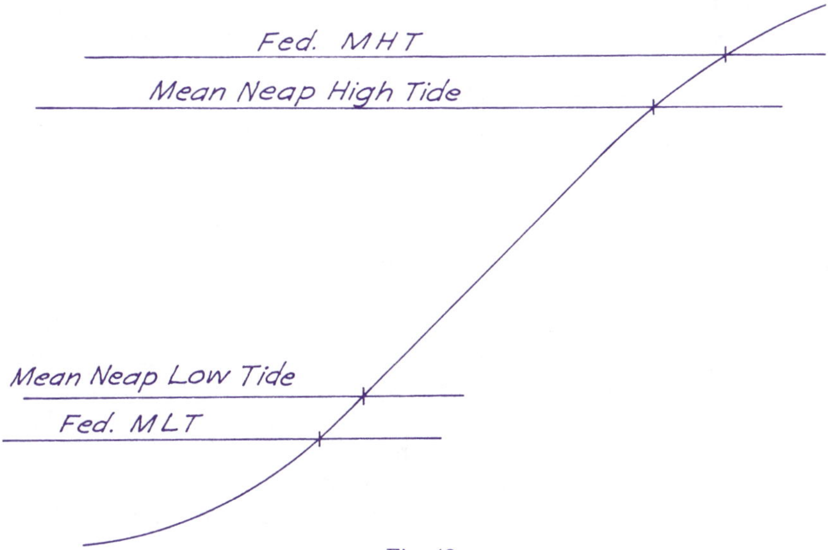

Fig. 42

[65]Manual of Instructions for the Survey of the Public Lands of the United States.
[66]140 Pac. 1092 (1914) and 18 Cal. 11 (1861) respectively.

not been established any specific length of time over which such observations should be made regarding neap tide line determination.

This analysis of the delineation of "ordinary high water mark" is given here to give you neither a positive answer nor a complete one that is applicable to every tideland description, but to alert you about the problems to which you must raise questions.

Outside of the instructions for the survey of public lands, the format to be used for tideland and waterfront descriptions, and the surveys therefor, have not been clearly defined. One exception was the Federal Act of May 24, 1824 which provided that "whenever, in the opinion of the President, a departure from the ordinary method of surveying land on any river, lake, bayou or water course would promote the public interest, he may direct the surveyor-general in whose district such land is situated, ..., to cause the lands . . . . to be surveyed in tracts of two acres in width, fronting on any river, bayou, lake or watercourse and running back the depth of forty acres; . . . . shall be offered for sale entire . . . . on the same terms in all respects as other public lands of the United States."

Various states have provided in their codes for the disposition of lands under water within their jurisdiction, but the mechanics and the format concerning the surveying and describing of those areas is not delineated. Again, California is unique in the procedure that was used for the determination of tideland descriptions for patent.

When an application for a tideland patent was presented to the county surveyor of the county wherein the area to be patented was located, he followed the requested area as nearly as possible by a survey. The county surveyor then wrote a description of it and presented it with his field notes, a sketch and the number of acres included to the State Surveyor General for his approval. In this process, no attempt was made to determine the line of the mean high water mark, by either the federal or state criterion, and generally neither the submerged nor interior upland areas were delineated for the benefit of the patent or patentee. Consequently, the resultant description, with no ties to "mean high water mark" or other pertinent conditions, became in effect a "hard line" type of description; that is, it merely followed the county surveyor's footsteps with no regard to legal boundaries.

Figure 43 illustrates a sample of disparity between the lines of a tideland patent description and the true lines of ownership as limited by the (1) line of mean high water (sometimes labelled MHW or MHT for mean high tide), (2) line of mean low water (MLW), (3) submerged areas, and (4) interior uplands. As previously mentioned, title to the submerged lands below mean low water remains in the state, so when

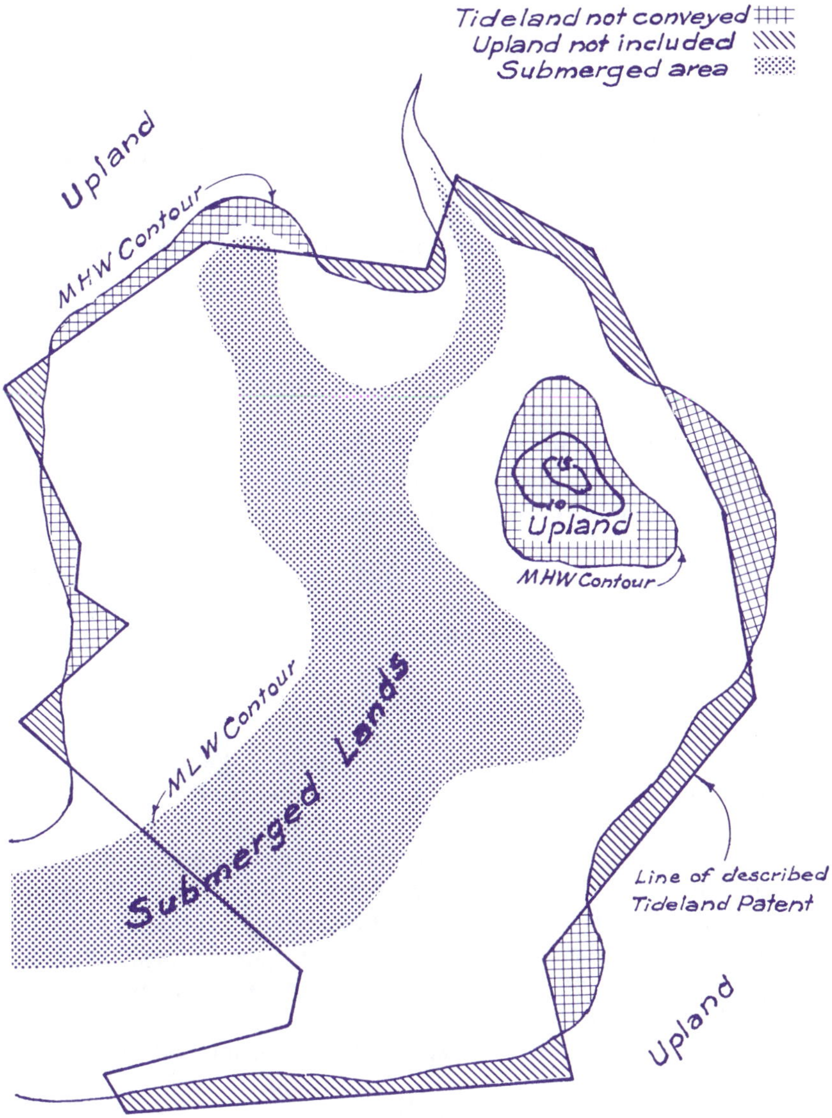

Fig. 43

that, along with the portions lying above the mean high water mark, are excepted from the described area of the tideland patent, the remaining net area is somewhat less.

From this it can be seen that it is necessary to review and determine the true boundary lines of upland title, as well as the submerged area, as compared with the described line because the grantor cannot convey that to which he has no title. It would be difficult to rewrite the description with any meaningful bearing and distances around the net area, but, to emphasize the unconveyed portions, EXCEPTIONS could be added, such as:

EXCEPT any portion thereof lying below Mean Low Water.

ALSO EXCEPT any portion thereof lying above Mean High Water.

Natural Boundary Line

Relative to the true position of any water boundary line is the need to determine where it was in its last *natural* state. This requires historic research of such items as floods, meteorological phases, the beginning of artificial conditions, surveys and surveyor's field notes. The solution is neither simple nor direct and should be coupled with legal counsel.

Any unnatural conditions resulting from dredging and filling, the establishment of groins, bulkheads or piers, etc., are not acceptable per se for ownership lines without suitable agreements or court case decisions.

Submerged Lands

The submerged lands and soils under tidewater held in trust for the people of the state are subject to improvements in the interest of commerce and navigation by either the federal government or the state, or a county or city holding such areas in trust for the state, and without the exercise of the power of eminent domain.[67] Therefore, "any evidence of title describing tidelands should contain an exception of rights and easements for commerce, navigation and fishery."[68] The Third District Court of Appeals in New Jersey did, however, hold that the federal rule under the River and Harbors Act contemplated that lines established for fill constituted abandonment of the rights of

[67]Newcomb v. City of Newport Beach, 7 Cal. 2d. 393, 401.

[68]SOURCES OF TITLE TO LAND IN CALIFORNIA, Title Insurance and Trust Co. 1965.

commerce, navigation and fishery (498 F.2d 597 [1974]). In areas where artificial lakes have been created, the rights of the public to fish therein, and, in some cases, the rights of commerce and navigation, too, have been established and held in the same manner as previously stated for other conditions.

Ownership Projected Into Tideland Areas

There are so many varied conditions in the configurations of tidelands and flats that many rules have been formulated in efforts to provide some semblance of fairness in the distribution of them to the adjoining upland owners, and the following are quotes of some of those rules as cited in CORPUS JURIS Vol. 9: "The rule for the division of flats is to run perpendicularly to the shore line from the point of division at high water mark to low water mark. (Conn.)" "Each proprietor of original tracts takes between the lines of his new frontage on the water course measured back to the old frontage. Course or direction of said lines is of no consequence. The extent of old frontage on the water course determines the extent of the new frontage. (La.)" "The intention of the ordinance of 1647 was, 'if practicable, to give to every proprietor the flats in front of his upland of equal width with his lot at high water mark.' The direction of the side lines of the flats is not governed by that of the side lines of the upland. Where there is no cove or headland, a straight line is to be drawn according to the general course of the shore at high water and the side lines of the lots are to be extended at right angles with the shore line. Around a headland the lines dividing the flats must diverge toward low water mark. (Mass.)" "A riparian owner when he extends his shore front, must, if the high water line is substantially straight, so extend his side lines as to make them rectangular with such high water line. (N.J.)" "In the absence of any agreement or facts working an estoppel, the line will be drawn from the termination of the line on the old shore to meet the harbor line perpendicularly. (R.I.)" "The mode of ascertaining the side lines of water lots from the upland to low water mark under the colonial ordinance of 1641, where they have not been otherwise settled by the parties, is to draw a base line from one corner of each lot to the other, at the margin of the upland, and to run a line from each of these corners at right angles with such base line to low water mark. If the line of the shore is straight, the side lines of the lots thus drawn to low water mark will be identical; but if by reason of the curvature of the shore they either diverge from, or conflict with, each other the land inclosed by both lines, or excluded as the case may be, is to be equally divided between the adjoining proprietors. (Me.)" In other words, the median line between the two conditions is the answer.

The last rule stated, having been considered such an equitable approach, has been used in many areas other than Maine. Its application in convex, straight and concave segments is shown in Figure 44a. Where surveyed meander lines exist, the boundary line in the tideland area is drawn at right angles from the meander line at its intersection with the upland boundary line.

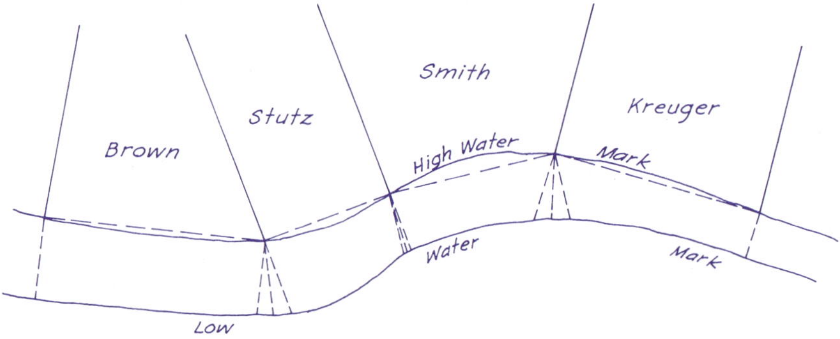

Fig. 44a

Boundaries of Stream, River, etc.

The thread of the stream is defined as a median line halfway between opposite shore lines measured neither in a dry condition nor a swollen condition but at an average or natural stage of existence.

Again, you must make very critical analysis of the wording in the descriptions.

If the deed specifically states,

> "to a point; thence northeasterly 12-½ rods to a point; thence southerly. . . . . . (etc.)"

with no tie to the creek or stream or river, then title does not carry to or into the creek, stream or river.[69] If, on the other hand, a call ties to the "bank of the creek, thence up the creek" with its meanders, fixes the boundary of the land at the margin of the creek, not at the middle thereof.[70] Or, in the instance of land described as "beginning at a

[69]Lynch v. Troxell, 207 Pa. 162,56 A 413, Schell v. City of Jefferson, 357 Mo. 1020, 212 SW 2d. 430 (1948).

[70]Fleming v. Kenney, 4 J.J. Marsh (Ky.) 155, 157.

stake standing on the bank or edge" of a certain lake and continuing from point to point on the "westerly bank or edge of said lake," it was held that the words "edge" and "margin" are synonymous and the words "bank or edge" were intended to mean the margin of the lake or edge of water and thereby leaving no strip of ungranted land between the edge of the lake and the top of the bank.[71] Also, in the use of water courses for a boundary, it is held that when reference is made to a river, stream or creek, the line of ownership will follow the twists and turns thereof and not a straight line between one end and the other, unless specifically so stated. One example read, "to the south bank of said river, thence up the same. . . .to the beginning,"[72] held that it meant to follow the meanders of the south bank.

In the case of apportioning the beds of non-navigable rivers, the MANUAL OF SURVEYING INSTRUCTIONS (1973) in Section 7-58 discusses the instructions for surveying partition lines as found in 50 L.D. 216 (1923) and the adaptation of the rule outlined in the case of Johnston v. Jones, 66 U.S. 117 (1861). This method is shown in Figure 44b.

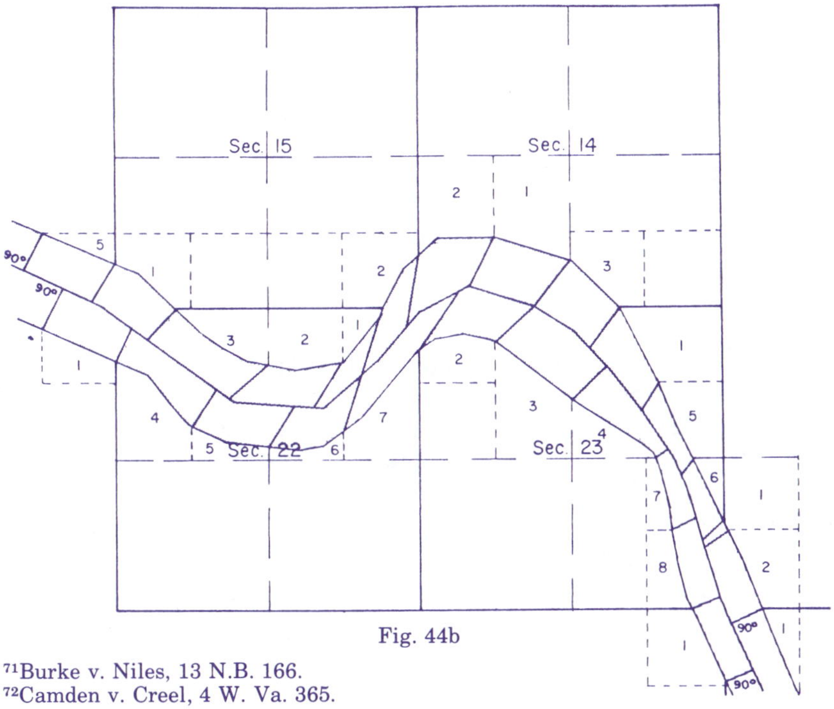

Fig. 44b

[71]Burke v. Niles, 13 N.B. 166.
[72]Camden v. Creel, 4 W. Va. 365.

An alternate method of apportioning the bed of a non-navigable river is discussed under Section 7-59 of the Manual and shown in Figure 44c.

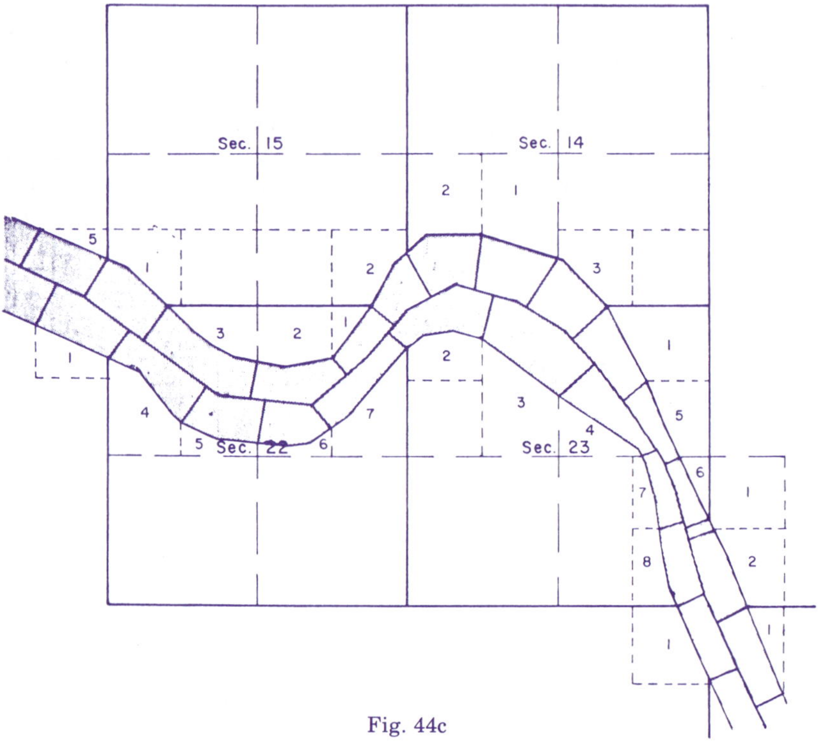

Fig. 44c

Outlines of ownership in *nonnavigable* rivers or lakes are normally not described by metes and bounds because the variety of circumstances excludes the adherence to a rigid rule. The form to use would be the description of the upland parcel plus, "Together with that portion of the adjoining river (or lake) which would pass with a conveyance of said land."

Once more, you are admonished to search the rules of land, both federal and state, about which you are concerned and review court cases concerned therewith prior to making final decision, either on the interpretation of the description before you, or as to the exact wording you will use in rewritting or creating a description.

Boundaries of Artificial Water Areas

In regard to ownership under and adjoining an artificial body of water such as a reservoir astraddle a dammed-up river, the boundaries are held as they existed in the original natural condition. See Figure 45a. If the river was navigable, the normal boundary would be the ordinary low water mark on each side of the river. Contrariwise, if the stream was non-navigable, title would carry to the thread of the stream.

In Texas, however, by the case of Strayhorn v. Jones,[73] a conveyance along the stream carries with it one-half of the bed whether it is navigable or not, subject to the state's rights in the bed. Similarly, in Michigan, a conveyance bounded by a stream, navigable or not, carries to the thread of the stream. This is not true in all states, though, and the status of navigability must be determined from the federal grants and statutes at the time of entering statehood and the subsequent statutes and court cases of that state. In some grants by Spain and Mexico, the entire river within the grant boundary was included. This should be researched in each case.

In fresh water, the ordinary high water mark is established by evidence of vegetation along the banks and the distinct marking of the soil by the continuous periodic action of the water to the extent that it makes it unfit for agricultural purposes.[74] It "is to be determined not from human records, but from the records which the river makes itself."[75]

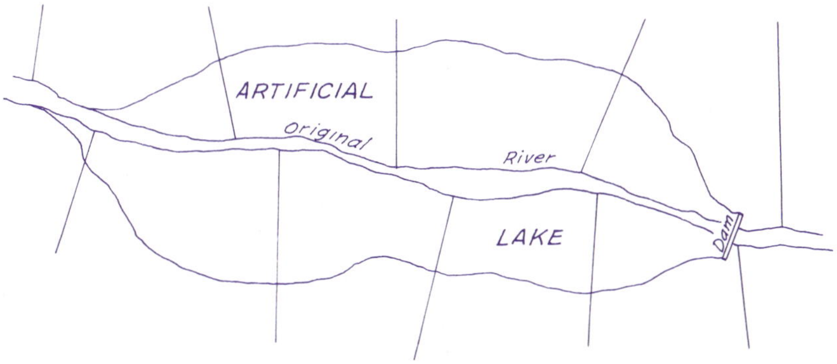

Fig. 45a

[73]289 S. W. 2d. 321, 330 (1956).
[74]Carpenter v. Board of Com'rs. of Hennepin County, 56 Minn. 513, 58 NW 295.
[75]Park Commissioners v. Taylor, 133 Iowa 453, 464, 9 C.J.

Lake Boundaries

Littoral lines of ownership apply to lakeshore boundaries as in the case of the non-tidal navigable lake shown in Figure 40. The limit of a land boundary bordering a natural pond or lake of which the water capacity varies with the seasons (not with any tide) goes to the lowest water mark,[76] that is, the water line of the lowest ordinary stage and not of a dry season, and free from natural disturbances.[77] However, whether the pond is natural or becomes artificial, either by depression from lowering the outlet or raised by a dam, the boundary limit is where it existed at the time of conveyance.[78] Furthermore, if the pond is returned to its original condition, the boundary will revert to its original low water mark.[79]

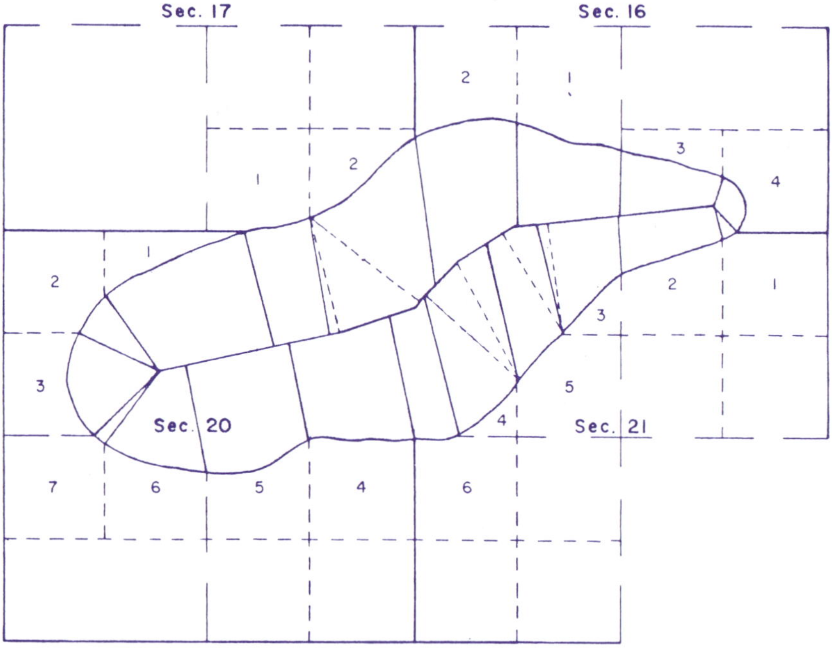

Fig. 45b

[76]Paine v. Woods, 108 Mass. 160, Wood v. Kelley & Noyes, 30 Me. 47, City of Boston v. Richardson, 105 Mass. 351.
[77]Slauson v. Goodrich Transp. Co., 94 Wis. 642, 69 NW 990.
[78]Bradley v. Rice, 13 Me. 198, 29 Am. D. 501.
[79]Waterman v. Johnson, 13 Pick. (Mass.), 261, 30 Mass. 261.

In the case of apportioning the beds of non-navigable lakes, the courts have held that a generally round one is cut like a pie with lines from the corners of properties meeting at a point in the center. The illustration in Figure 45b is from Section 7-61 in the MANUAL OF SURVEYING INSTRUCTIONS (1973) showing the apportionment of the bed of an irregular shaped nonnavigable lake with certain alternatives in dashed lines.

Another aspect of lakeshore boundaries is applicable to the Great Lakes in which the lands under inland navigable waters are held in trust by several bordering states.[80] This was stated as early as 1892 when, in the case of Illinois Central Railroad Co. v. Illinois,[81] the Supreme Court upheld the right of the state to repeal a previous grant of submerged lands to the railroad because the state held them in trust for the people and could not abdicate that trust by giving an irrevocable grant.

Accretion, Alluvion and Reliction

ACCRETION is the act of growing to a thing; usually applied to the gradual and imperceptible accumulation of land by natural causes, as out of the sea or a river. ALLUVION (Black does not list Alluvium but Webster shows Alluvion = Alluvium) is that increase of the earth on a shore or bank of a stream or the sea, by the force of the water, as by a current or by waves, which is so gradual that no one can judge how much is added at each moment of time. Rules differ in different areas, though, regarding the rights of riparian and littoral owners to alluvion. "Accretion" denotes the act; however, the terms are frequently used synonymously. Katz v. Patterson, 135 Or. 449, 296 Pac. 54, 55. "The common law and federal law of accretion, unlike the California rule, do not require that the alluvion be deposited through 'natural causes' in order to belong to the riparian or littoral owner. As stated in the 1965 federal decision of *Beaver v. United States* [350 F. 2d 4-9th Cir. 1965] in connection with riparian upland of the United States. 'The erection of artificial structures does not alter the application of the common law doctrine of accretion (alluvion formed by slow and imperceptible degrees) unless the structures are erected for the purpose of causing the accretion.' "[82] RELICTION is an increase of

[80]Confirmed in Public Law 31, see note 48 supra.
[81]146 U. S. 387.
[82]TITLES TO LAND IN THE COASTAL ZONE, by Thomas E. McKnight in California State Bar Journal, Sept./Oct. 1972.

the land by the permanent withdrawal or retrocession of the sea or a river.[83]

Where property is held in status with the water line as a boundary, whether it be by metes and bounds with a tie to the water or a lot in a tract that shows a water line as a boundary, any gradual accretion, reliction or erosion carries the title boundary with it. This applies to both fresh water and tidewater.

The distribution of that land which is added to water boundary property follows several different rules and, again, you must discover which one is applicable to your particular case. The allotment of the land so added to fresh water boundaries adheres to the format of proportioning and follows one of these rules:

1. Proportionate acreage based upon a direct relationship to the area held by each respective owner at the time of conveyance.
2. Proportionate shore line distance, whether it be straight, convex or concave, based upon a direct relationship to the frontage held by each respective owner at the time of conveyance.
3. Proportionate thread-of-the-stream measurement (similar to 2 above) based upon its direct relationship to the frontage held by each respective owner at the time of conveyance.
4. Around a generally circular lake, the center is established first and lines are drawn from it to the sidelines of each property where they intersect the original peripheral boundary. This is also known as the pie method.
5. Where a lake is elongated, and perhaps also irregular, a combination of the thread-of-the-stream and the pie method is used; in the main portion, lines are drawn from the property corners perpendicular to the thread-of-the-lake, similar to thread-of-the-stream in that it is a line midway between opposite shore lines taken the long way of the lake, while the concave ends are cut into sectors by the pie method.

Avulsion and Erosion

AVULSION is the removal of a considerable quantity of soil from the land of one man, and its deposit upon or annexation to the land of another, suddenly and by perceptible action of water. Because the change is not gradual and imperceptible, the ownership does not change and follow the new water lines; in other words, if a river had

[83]Black's L. D.

formed an ox bow but during a flood it jumped across and cut it off, the owner would retain the part on the other side of the new river bed: the original lines of title remain as they existed prior to the avulsion.

EROSION is "The gradual eating away of the soil by the operation of currents or tides. Distinguished from *submergence* which is the disappearance of the soil under water and the formation of a navigable body over it. Mulry v. Norton, 100 N.Y. 433, 3 N.E. 584, 53 Amer. Rep. 206, State of Arkansas v. State of Tennessee, 246 U.S. 158, 38 S Ct. 301, 304, 62 L. Ed. 638, L.R.A. 1918 D. 258."[84]

Clark on Surveying and Boundaries cites Payne v. Hall and Webber v. Axtell,[85] "Where the lands of a riparian owner are removed by the gradual process of erosion by the river, the land being no longer capable of identification, but having been carried away entirely, and the river occupies the identical space formerly occupied by the lands of the riparian owner, the title to the land so occupied by the bed of the river passes from the owner to the state."

From these definitions, it is seen that accretion and alluvion are considered synonymous in respect to the addition of earth to earth as a result of gradual current or wave action. Contrariwise, reliction *leaves* an addition of land area by the gradual permanent withdrawal of water. Avulsion is different from these actions in that it is sudden and recognizable in a short period of time. And erosion is the gradual eating away of land as a result of current or wave action.

Conclusion

When the question is raised concerning the physical location of any water boundary, one must always start with the qualification which is determined by the nature of the subject; i.e., tidal or non-tidal, navigable or non-navigable, etc. Following this must be the determination of the position of the water line in its last natural state which may require for verification its relative position at the time of the issue of the patent with attendant consideration of any accretion, reliction or erosion and avulsion, if any. Further observation and study must be made of any artificial conditions affecting the prior natural status. It has been said that, depending upon the direction and force of currents, such things as groins, levees and piers may affect water lines as far as ten miles away.

[84]Black's L. D.

[85]192 Ia. 780, 185 NW 912 and 94 Minn. 375, 102 NW 915, 6 L.R.A. (NS) 194 respectively.

Therefore, the intricacies involved in water boundaries require painstaking and exhaustive research and analysis before any decisions can be considered. Writing descriptions of such boundaries should be done only after the scrivener has obtained a thorough acquaintance of the existing facts, historical background, state and federal statutes concerned therewith, court cases and opinions relative to the conditions at hand.

It is an exceptionally rare situation where a description of the line of the water itself will be required. Usually a meander line approximating the water's edge will be surveyed and your description will follow that meander line. In this case, you must be sure to designate it as such so that the title will go automatically to whatever line of water legally prevails under the circumstances involved.

Where historical conditions are so vague as to be indecisive, or so involed with such multiplicity of facts that there is no clear cut answer evident, the State Lands Commission (or whatever body politic has the authority and responsibility for the dispensation of State lands) may offer to establish and agree to an independent arbitrary agreement line between itself and the upland owner. If you are dealing with the establishment of such an agreement line, along the ocean for instance, be certain that the wording in the description does NOT refer to, "mean high tide line" because that statement inserted will compel the agreement line to revert to the exact problem from which it is intended to free the land. Make the agreement stand on a clearly described line tied to well established and acceptable monuments and controls of record.

PROBLEMS

1. What are the upper and lower limits of tide land?
2. What is the federal time interval for determining the vertical difference of the limits of tide land?
3. Is a meander line by definition a boundary of ownership?
4. Can an individual own "submerged" land?
5. Does ownership of land adjoining a man-made reservoir carry title under the water?
6. What are the five rules stated for the distribution of land added to fresh water areas by accretion, alluvion or reliction?
7. Does the limit of use of riparian rights of a water boundary owner coincide with the boundaries of his fee title?

Chapter 7

Analysis and Interpretation of Descriptions

Sufficiency

One of the first details to look for in analyzing any description is the matter of sufficiency. One of the primary questions is whether there are enough references to documents in the public office of records, including maps, and are they, in turn, sufficient? It is possible that the reference material has shortcomings, and if it is not satisfactory, then its value in the immediate description is questionable and the effect can be upsetting. A lack of, or insufficient, description of monuments can also create indefiniteness. Ambiguity, whether it is in the distance given, or in the tie given, or in the material given for reference, or the absence of facts, all helps to diminish or negate the value of the description. Insuffícency can even void a conveyance.

An example of an indefinite tie is:

> "to the west line of the grantor's land."

The grantor may own more than one piece of land and you would not know which west line it is nor where it is in relation to your objective.

Sometimes descriptions are written which are incomprehensible without the help of a map or deed. For example:

> "Beginning at the southerly terminus of course number 21 of Lot 10, said point of beginning being the most westerly corner of land described in deed to R. Brothers . . . ."

From this you do not know in what part of Lot 10 the point of beginning is without the additional information from the deed or its references.

Another arrangement of this description which would give you a mental picture of the situation without having to dig around for a map would be thus:

> "Beginning at the northerly terminus of course number 21 which is the northwest corner of Lot 10; thence south 15° 45′ 00″ East 282.93 feet along the westerly line of said lot and course number 21 to the southerly terminus thereof, being also the most westerly corner of the land described in deed to R. Brothers . . . ."

In this way, by describing the fact that course number 21 is also the westerly line of Lot 10 and the north end of that course is the northwest corner of the lot and the south end of that course is also the most westerly corner of R. Brothers, you then, have the mental picture of the juxtaposition of these parts. This type of analyzing the parts first should be used in writing descriptions so others who follow will have a better and easier comprehension of what you are trying to express.

Another point of insufficiency which makes for poor understanding is the improper positioning of those items which control the location of the particular piece of land in relation to the description as a whole. Sometimes a description will wander all over the area before it describes or pinpoints the actual ties and the pertinent items which control that boundary. Consider the following:

> "Beginning at the easterly terminus of that certain course having a length of 238.56 feet in the northerly boundary of the land described in deed to Lesarm Seego and wife recorded in Book 2390, Page 111, Official Records; thence along said boundary North 89° 22′ 36″ West 238.56 feet to the westerly terminus of said course; thence South 0° 06′ 07″ West 648.78 feet to the northerly line of land described in deed to Billy Hill and wife recorded in Book 1781, Page 23, Official Records; thence in a generally southeasterly direction along the last mentioned northerly line . . . ."

From all of this you still do not know where you are without getting the deeds and maps, if any, and drawing all the parts. The preferable way would be thus:

> "That portion of Farm Lot 201 in the Rancho Parte de San Pascuale in the county of Magoo, State of New Mexico, as per map recorded in Book 12, Page 10, of Maps in the office of the County Recorder of said County described as follows:
>
> Beginning at the intersection of the south line of 2nd Street with the west line of View Street as shown on said map; thence South 0° 02′ 38″ East 492.61 feet along the said west line to the true

> point of beginning at the northeast corner of the land described in deed to Lesarm Seego and wife recorded in Book 2390, Page 111, Official Records; thence along the north line of said Seego land North 89° 22′ 36″ West 238.56 feet to the northwest corner thereof; thence . . . . . etc., etc."

In the last form it is possible to visualize the development of that parcel from a known point. Having to search for ties to monuments or adjoiners or anything by which to pin down a line or a corner proves insufficiency of the description. The lack of sufficiency could invalidate the title.

Conflicting Elements

A detective's skills are often helpful and sometimes an absolute necessity in solving a problem description.* Consider the following taken from a deed:

> "Commencing at a point 54 feet southeasterly parallel with and along the easterly line of Lot 3 Block 7 of the Nutwood Place in the city of Orange; thence westerly parallel with the southerly line of said Lot 3 to a point 161.98 feet to a point 20.58 feet westerly from the easterly line of Grand Street in said city of Orange; thence South 56.80 feet to a point; thence North 75° 01′ East 56.25 feet to a point; thence North 69° 11′ East 113.82 feet to a point; thence North 51° 36′ 30″ East 165 feet to a point on the westerly line of River Avenue extended southeasterly; thence North 51° 32′ West 85.58 feet along the line of River Avenue to the southeasterly corner of Lot 4 in said Block 7 in the said Nutwood Place; thence westerly along the southerly line of said Lot 4, 115.20 feet to the southwesterly corner of said Lot 4; thence southeasterly perpendicular to the said southerly line of said Lot 4, parallel with and extending the westerly line of said Lot 4, 54 feet to the point of beginning."

When the point of commencement is intended to be on the southeasterly prolongation of a line, it is NOT "parallel with and along the easterly line of Lot 3" because it is outside of Lot 3. When the lot is on a 40° diagonal, the side line is NOT "westerly along the southerly line;" it would be "southwesterly along the southeasterly line." When

*Taken from DESCRIPTIONS — DEBATABLE OR DEFINITIVE? Gurdon H. Wattles, Journal of the SURVEYING AND MAPPING DIVISION, Proceedings of the American Society of Civil Engineers, January 1967.

there are no bearings shown on the lot lines on the tract map, but bearings were derived somehow by someone for use in the description, then they should be related to some former line of control, i.e., a street line, and NOT drop out of a clear sky into the middle of the description and onto a new line having no qualification. Perhaps it should have been in this case, "parallel with the southerly prolongation of said Grand Street." When there are no bearings on the lot lines in said tract, it canNOT be assumed that "southeasterly perpendicular to the said southerly line of said Lot 4, parallel with and extending the westerly line of said Lot 4" is either a line at 90° or parallel, or both. Obviously, the facts have been considerably muddled.

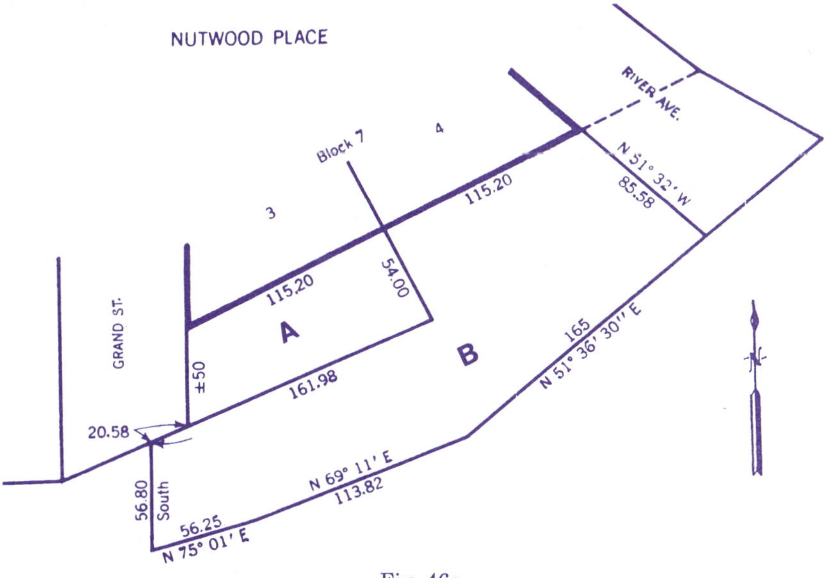

Fig. 46a

Without going into the *modus operandi* of determining the intended outline of the land to be included within the above aggregation of words, a proper description of Parcel B as shown in Figure 46a might be:

The portion of Lot 12 in Block D of the Chapman Tract, in the city of Orange, county of Orange, state of California as per map recorded in book 102 page 15 of Miscellaneous Maps in the office of the county recorder of said county, described as follows: Beginning at the most easterly corner of Lot 3 in Block 7 of NUT-

WOOD PLACE in said city as per map recorded in book 4 page 70 of said Miscellaneous Maps; thence along the prolongation of the northeasterly line of said Lot 3 southeasterly 54.00 feet; thence parallel with the southeasterly line of said Lot 3 southwesterly 161.98 feet to a point distant southwesterly 20.58 feet measured along said line from the East line of Grand Street shown on said map of NUTWOOD PLACE; thence parallel with the southerly prolongation of said East line, SOUTH 56.80 feet; thence North 75° 01′ East 56.25 feet; thence North 69° 11′ East 113.82 feet; thence North 51° 36′ 30″ East 165 feet to the southeasterly prolongation of the southwesterly line of River Avenue shown on said map; thence along said prolongation North 51° 32′ West 85.58 feet to the most easterly corner of Lot 4 in said Block 7; thence southwesterly 115.20 feet to the point of beginning.

Concerning the use of the word SOUTH for a bearing with no prior relationship and also using it in the capacity of a base of bearing as it is here, the courts have held that this would be represented properly by "true" SOUTH established from an astronomical determination. The position of the lot lines within the Tract must be developed internally and separately from the lines having bearings based on Grand Street.

This is a case where the revised description becomes a little longer than the original, but it is clear and definitive. The correct choice of words is the prime factor in composing a description. There may be too many words which can easily fall into paradoxical juxtapositions or there may be too few words leaving gaps of information which thereby fail to give a complete picture and may fail to convey full or proper title.

Sometimes the scrivener lacks understanding, and this is reflected in the way he writes. The following description is a good example of this:

"All of the northerly half of all that certain lot, piece or parcel of land situate, lying and being in the Rancho Santiago de Santa Ana, County of Orange, State of California, bounded and particularly described as follows, to wit:

Commencing at the northwest corner of Lot L of the Chapman Tract, as surveyed by E. T. Wright April 13, 1881, map of said survey being recorded in book 4, page 404, Miscellaneous Records of Los Angeles County, California. Running thence South 25′ East 14 chains and 78 links; thence North 89° 15′ East 10

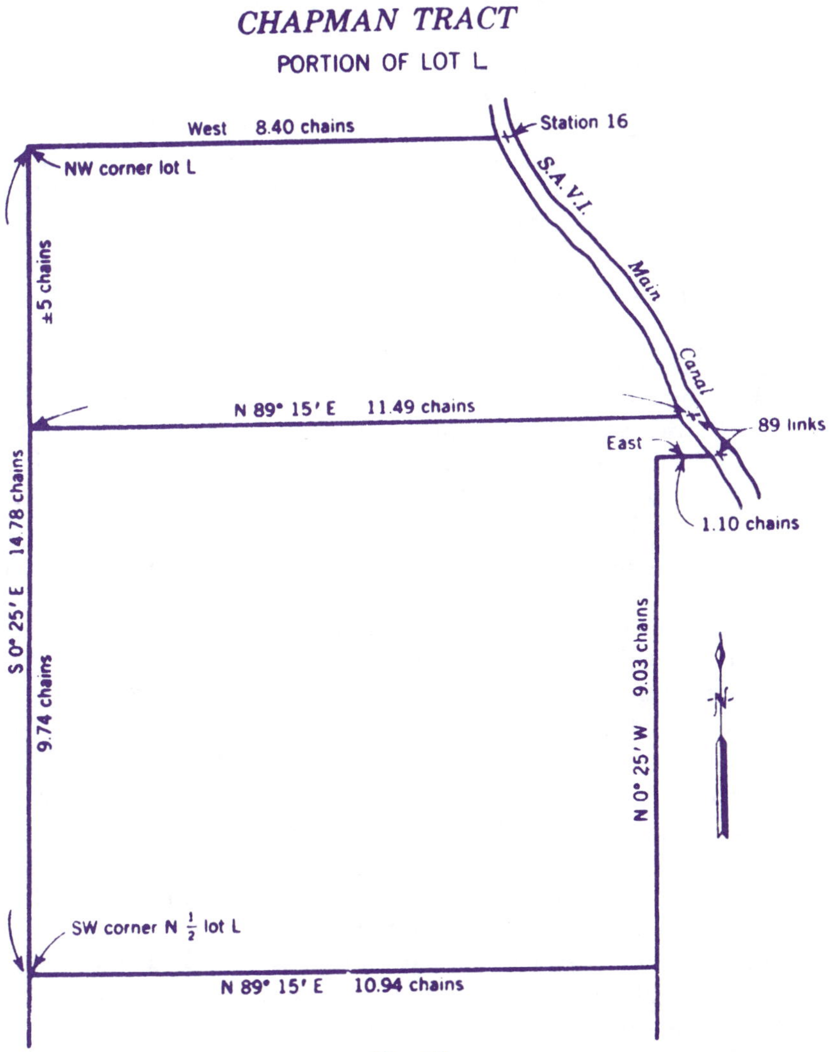

Fig. 46b

chains 94 links; thence North parallel with said first line 9 chains 3 links, thence East 1 chain 10 links to center of Main Canal of S.A.V.I. Co.; thence northwesterly along the center of said Canal to Station 16 of survey of said E. T. Wright; thence West 8 chains 40 links to the place of beginning.

> Being 16 acres of land less and excepting the following described land sold off of said above described land:
> Commencing at the southwest corner of the North ½ of Lot "L" of the Chapman Tract, running thence North 25′ West, 9 chains 74 links, along the West line of said Lot "L"; thence North 89° 15′ East 11 chains 49 links to the center of the Main Canal of the S.A.V.I. Co.; thence along the center of said Canal 89 links; thence West 110 links; thence South 25′ East 903 links; thence South 89° 15′ West 1094 links to place of beginning."

From an analysis of extrinsic information, the following describes the land correctly as shown in Figure 46b:

> That portion of the North half of Lot L in the Chapman Tract in the Rancho Santiago de Santa Ana, County of Orange, State of California, as per map of survey by E. T. Wright, April 13, 1881, recorded in book 4 page 404 of Miscellaneous Records in the office of the county recorder of Los Angeles County, California, described as follows:
>
> Beginning at the northwest corner of said lot; thence along the westerly line of said lot South 0° 25′ East 5 chains, more or less, to a point North 0° 25′ West 9.74 chains from the southwest corner of said North half; thence North 89° 15′ East 11.49 chains to the center of the Main Canal of the S.A.V.I. Co.; thence northwesterly along the center of said Canal to the northerly line of said lot; thence West 8.40 chains to the point of beginning.

The brevity, simplicity, and completeness of the second description is easier to understand, has less chance of error, and is less costly to record.

Perhaps the old timers should not be blamed too much for their verbose language considering that they were paid only $0.02 per word for writing a description. After all, they had to earn a living!

Sometimes statements are made in a description which appear to conflict either with other statements therein or with citations in description of adjoining properties. In order to have a method of discernment in the relativity of values when comparing descriptions, there is a generally recognized and conceded order of importance assigned to the principal elements as discussed in Chapter 4. Conflicting facts are sometimes called "conflicting elements" when, in reality, although some do conflict, generally each one has its part in supporting the whole story told by the description. The different ratings of the elements delineate their status and establish the stairsteps of values by which you can decipher and hence interpret the description.

Adjoiners

While the citation of a definite monument carries a high priority, still a call to an adjoiner is in itself a call to a monument and where it is certain it is held in high dignity.[86] "Calling for a well known line of another tract, denotes the intention of the party, with equal strength, to calling for a natural boundary, so long as that line can be proved."[87] "A call for the line of another tract of land is a 'natural boundary.' "[88] This is of course based on the adjoining deed having priority of time over the description wherein it is called for and the fact that the call to the line of a proven (established) adjoining tract is more certain than mere bearing and distance.[89]

If lines and/or corners are not called for as boundaries in a conveyance, they are not controlling.[90] If the reference to adjoiners is merely incidental and the context does not make the tie positive, the call must be considered as only locative.[91]

The rule for holding to a recited adjoiner may be abrogated if it is incapable of definite ascertainment.[92]

Artificial Things

Regarding control by artificial monuments and marks, Corpus Juris states that, "After natural and permanent objects, artificial monuments and marks, unless clearly erroneous, control other and inconsistent calls for boundaries,[93] provided the monuments are mentioned in the deed. The rule has no application to a description in a deed which does not mention any monuments."[94] "In case of conflict

[86]Miller v. Grunsky, 141 Cal. 441, 66 P 858, 75 P 48.
[87]Cherry v. Slade, 7 N.C. 82, 90.
[88]Graybeal v. Powers, 76 N.C. 66.
[89]Love v. Jones, (Tex. Civ. A) 138 SW 1128.
[90]Mathews v. Thatcher, 33 Tex. Civ. A, 133, 76 SW 61.
[91]Waters v. Dennis Simmons Lumber Co., 154 N.C. 232, 70 SE 284.
[92]Brolaskey v. McClain, 61 Pa. 146, Boon v. Hunter, 62 Tex. 582, Brown v. House, 116 N.C. 859, Day v. Wilder, 47 Vt. 583, Hill v. McConnell, 106 Md. 574, 68 Atl. 199, Starke v. Johnson, 2 Mill. Const. 9, 4 So. C. 1.
[93]Resurrection Gold Min. Co. v. Fortune Gold Min. Co., 129 Fed. 668, Watkins v. King, 118 Fed. 524, Ulman v. Clark, 100 Fed. 180, U.S. v. Murray, 41 Fed. 862.
[94]Talbot v. Smith, 56 Or. 117, 107 P 480, 108 P. 125.

between calls, those for natural or permanent objects will control those for artificial monuments, marks or lines,[95] and especially artificial monuments at remote distances,[96] unless it clearly appears that the artificial monument is the more reliable of the two."[97]

Maps, Plats and Field Notes

It is common practice to refer to maps, plats and/or field notes in descriptions and it has been held that such reference includes facts shown on that reference and legally incorporates them into the description,[98] and furthermore, they are deemed to possess the same status, as such, as monuments.[99] In the event that such references conflict with other calls, then the application of the rules of construction (see Chapter 4) come into play and the true intent of the parties to the transaction is to be considered giving full benefit to those calls which are most definite and certain.[1]

It has been held in rare situations that if parties to a transaction had common knowledge and agreed understanding of the facts based upon a map mutually recognized, even though it was unrecorded, the map would be the governing factor.

As to the relative value between maps (plats) and field notes, there are two approaches to making a decision as to which one will prevail. Upholding the map in Haley v. Martin,[2] the court said, "It is questionable if any man buying a recorded map would bother about field notes." Conversely, the field notes were upheld in Harrington v. Boehmer[3] where the court said, "Where there is a discrepancy between field notes and a plat, the latter being made from the former, and the former being the better evidence as to where the line was run in the field, the plat must give way to the field notes." The latter statement is probably more applicable to government surveys where the official field notes are commonly used in connection with the plat

[95]Brown v. Huger, 21 How. 305, 16 L. ed. 125, Newsom v. Pryor, 7 Wheat 7, 5 L. ed. 382, McIver v. Walker, 4 Wheat 444, 4 L. ed. 611.
[96]Kimball v. McKee, 149 Cal. 435, 86 P 1089.
[97]Whitehead v. Ragan, 106 Mo. 231, 17 SW 307, Seaman v. Hogeboom, 21 Barb. (NY) 398, Bolton v. Lann, 16 Tex. 96, Hurley v. Morgan, 18 N.C. 425.
[98]DeLancey v. Wellbrock, 113 Fed. 103, Dallum v. Breckenridge, 3 Tenn. (Cooke) 152, Owsley v. Johnson, 95 Minn. 168, 103 NW 903.
[99]Wilson v. Chicago Lumber etc. Co., 143 Fed. 705, 74 CCA 529.
[1]Parker v. Kane, 22 How. 1, 16 L. ed. 286.
[2]85 Miss. 698, 38 S 99.
[3]134 Cal. 196, 66 P 214.

and are referenced on it. In contrast to public land subdivisions such as township plats, refererences to private subdivision maps or plats seldom, if ever, mention the field notes so the former quote from Haley v. Martin would be a reasonable application.

Metes and Bounds

In the matter of conflict between reference to the facts shown on a map and a metes and bounds description of the same area, there are also two approaches. In Goldsmith v. Fillman,[4] the court held that the metes and bounds description around a block of lots which disagreed with the boundary of that block shown on a map referred to in the same deed would give way to the measurements shown on the plan. Also in Chadwick v. Carson,[5] the court held that where a metes and bounds description correctly described an area which was incorrectly referenced by government numbers, the former controls. Contrariwise, other cases held that a description of numbered lots shown on a map will control over a metes and bounds description.[6] Because of these dual decisions, you must be extensive in your analysis of similar situations, research other cases comparable to your instant problem and use very considered judgment before reaching your conclusion.

Again with metes and bounds, you may have some conflicting items but otherwise, it is a matter of discernment in assigning the correct values to the various parts. As always, specific definitiveness holds over generalities and ambiguities and the rules of construction are to be applied as appropriate.

General Descriptions

When a description is defective, the use of a general description can be used to assist with interpreting the intent but it may not control or override.[7] If there are obvious errors in a metes and bounds description, it has been held that one of generality, referring to monuments, will prevail. Sometimes these are found as a supplemental statement in or after the main body of the description and other times they appear as a completely separate description.

[4]33 Pa. Super. 44.
[5]78 Ala. 116.
[6]Davidson v. Arledge, 97 N.C. 172, Inter City Realty Co. v. Newman, 128 App. Div. 195, 112 NYS 481, Cook v. Hensler, 57 Wash. 392, 107 P 178.
[7]McFaddin v. Johnson, (Tex. Civ. A) 180 SW 306.

Other types of general descriptions include a blanket reference, a house number, an owner's name or "unsold" lots. In the case of a deed covering "all land of the grantor" with no specifics, other evidence may be necessary. With a conveyance referenced to a house number, parol evidence may be required to identify the boundaries of the land under question before it can be considered marketable and insurable.[8] If the deed only states, "The south half of my lot in ___ City." and the grantor owns only one lot, it can be considered certain. A description of "all unsold lots" or "all unconveyed lots" introduces a possible question of whether the deeds of any of the "sold" lots may not have been recorded, thus leaving the buyer uninformed via the normal channel of constructive notice.

Natural or Permanent Things

Natural monuments are commonly given preference over other calls because the likelihood of mistaken identity is less. Mistakes may be made in measuring bearings and distances, but to define and cite a certain monument which can be recognized in later times leaves no question as to the control intended.[9] A condition of conflict can exist, however, if the use of that monument creates a condition of aggrandizement which of course is not permissible. "When they (natural monuments) are noted in the field notes as mere incidental calls in passing, their reliability is weakened, and sometimes rendered wholly worthless. Distances called for between corners to creeks or roads unless especially designated in such a manner as to show the intention to make them locative are not such, and will not ordinarily have precedence over a call for course and distance."[10]

Where two lines are described in two different deeds, each purporting to be the same boundary but one describes a monument and the other does not and that monument fits one of the lines, it has been held that that line with the monument was the one intended.[11]

Marked or Surveyed Lines

Lines established on the ground by survey with monuments set are held as the highest and best evidence of true location. Following in the footsteps of the surveyor is based on his following the orders of his

[8]Von Rohr v. Neely, 76 CA 2d 713, 173 P 2d 828.
[9]Howe v. Bass, 2 Mass. 380, 3 Am. D. 59.
[10]Jones v. Andrews, 72 Texas 5, 9 SW 170.
[11]McAdoo v. Sublett, 1 Humphr. (Tenn.) 105.

client. Conflict is created when there is mistranscription of the field notes or wrongly written descriptions or imposition on other areas of ownership. All of the above is based on the assumption that the work on the surveyed and marked line was done prior to or contemporaneously with the execution of the deed and in agreement with the understanding of the parties concerned. In contrast, though, it has been held that the rule does not apply where no reference to the lines of the survey is made in the deed.[12] In Hamilton v. Blackburn,[13] the fact was pointed up that lines cannot be constructed on the ground according to objects found presumably set by the surveyor but not called for in the description, even though it appears they were set in contemplation of a deed of conveyance[14] because the mere running of a line in contemplation of a deed does not bind the grantor since he would be free to change the line and the contract represented by the description.[15]

Ties

Normally a tie, or a call, given in a description to an adjoiner deed, a physical monument, a railroad right-of-way, etc., etc. means that whatever distance and/or bearing (depending on the wording given for the tie) is measured on the ground to that tie will hold.

There is another form of call, however, that results in a different application. If it is written thusly:

> ". . . . ; thence easterly 116.74 feet along the south line of said lot to a point which is also westerly 64.78 feet from the southeast corner of said lot; . . . . "

the position of that point on the south line of the lot is established by proration, NOT by holding the second call. If the words "which is also" were deleted, the point would not be established by proration but would be held at the 64.78-foot tie distance from the southeast corner.

MISTAKES

Mistakes occur in descriptions for various reasons, such as typographical errors, lines omitted, misreading of someone's writing or

[12]Goldman v. Hadley (Tex. Civ. A), 122 SW 282, Elliott v. Jefferson, 133 N.C. 207, 45 SE 558, 64 LRA 135.

[13]43 Tex. Civ. A. 153, 95 SW 1094.

[14]Elliott v. Jefferson, 133 N.C. 207, 45 SE 558, 64 LRA 135.

[15]Ritter Lumber Co. v. Montvale Lumber Co., 169 N.C. 80, 85 SE 438.

even errors in mathematics used in preparation of the description. Some of these are obvious, some may be deciphered, while others are evidenced by mathematical non-closure. The courts have held that if all of the surrounding information can be verified except for the one item which does not fit, then the description as a whole is not defeated and all evidence is used to support it. In other words, the mistake that has been proven to be such is recognized and thrown out in favor of

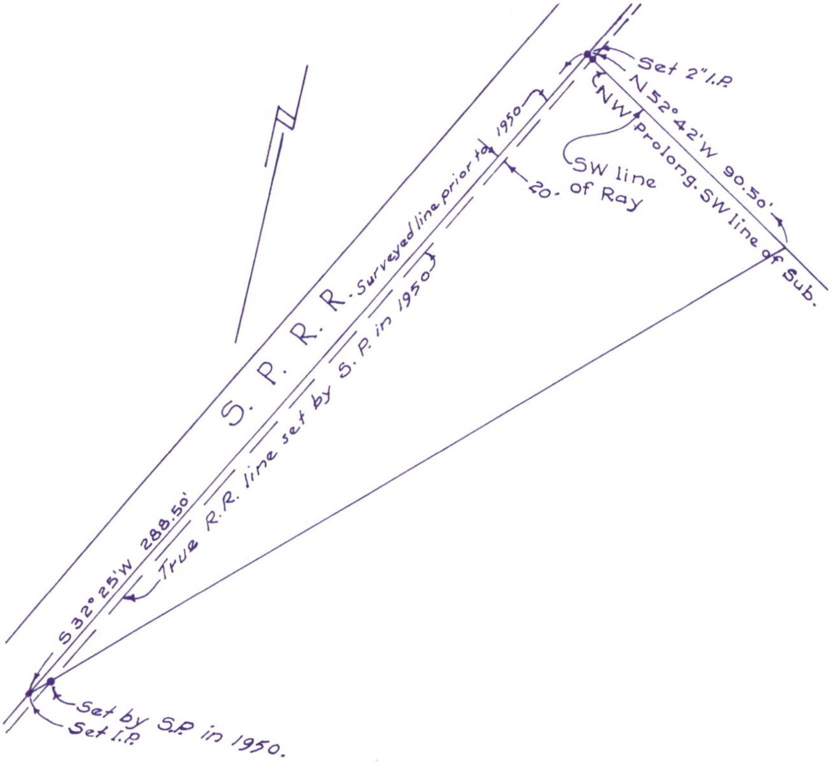

Fig. 46c

the remainder. A mistake which is obvious or one that can be determined by and through the process of elimination can be handled; however, when there is ambiguity and the mistake is not discernible, and in fact there are two or more possible solutions, then the mistakes must be recognized and other avenues of determination looked for.

Even in such cases, there may well be the necessity for a new description which is compatible to the surrounding facts and acceptable to the adjoining owners who are involved.

That which is most certain and definite will prevail over the less certain and indefinite. When the exact position of a boundary is indeterminate, that construction of the description is favored which gives proper effect to every call and which does least violence to the calls.[16]

Another form of mistake is when an incorrect joinder property is inadvertently used or a joinder which has been assumed to be satisfactory proves to be invalid. Several court cases have upheld and favored the use by mutual acceptance of monuments by adjoining owners in contradiction to the calls in the deeds of one neighbor to the other. Again, this is not the only rule to follow because the conclusion depends upon the circumstances involved so rather than try to outguess the court yourself, seek the counsel of your attorney or ask the title company what they will insure.

Indeterminate lines or corners may be a result of multiple incompatible ties. An example of this is the one shown in Figure 46c. The description was written:

> ". . . . ; thence north 52° 42′ west 90.50 feet along said prolongation to a set two inch iron pipe on the railroad right of way line; thence south 32° 25′ west 288.50 feet to an iron pipe on said railroad right of way line; thence northeasterly on a straight line to the point of beginning."

The map furnished by the surveyor showed the iron pipe to be on the boundary line of the railroad right of way as described. Subsequently, however, the railroad's own crew ran a survey several miles up and down the right of way, and by checking other monuments and lateral ties, established the true line some 20 feet southeasterly of the line between the two iron pipes set by the first surveyor.

In review of the superiority of calls, the rule states that monuments cited in a description have a very high priority in control so what happens when the monuments set and described do not conform with the correct position of the line which they were intended to represent?

The buyer of this triangular piece of land was on the ground when the railroad surveyors were doing their work and they informed him

[16]Stacy v. W. M. Ritter Lumber Co., 114 Va. 133, 75 SE 1038.

of the difference and the apparent encroachment of the iron pipes on his corners into the land of the right of way. The ultimate result was that the buyer went to court for determination. The court held that the pipes set by the first surveyor were focal and that the new owner could, on that basis, expect to be the holder of the 20 feet of the railroad right of way; however, since the grantor can only convey that which he owns and he did not own any part of the right of way, the 20 feet could not be included in the new ownership.

Because this description was insured by a title insurance company, they paid the damages awarded by the court; without title insurance, the private surveyor would have had to pay off.

Reverse Order

Another type of problem is created when the descriptions are rewritten in reverse order. Such a problem came to life when a survey based on a current recording of a deed showed that all points except two were out of position in their relationship between each call and the respective monuments on the ground. This did not appear to be proper and a search of the chain of title back to the turn of the century revealed that the original document contained the description written in the reverse order from the present day one. When that original wording was applied to the information shown on the survey, it immediately brought into focus the fact that all the points were OK within reason and that the error of some 8 feet was actually in that closing course. This indicated that the scrivener who wrote the first description on the first deed knew there was a problem and he put the error in the closing course where it belonged. Whoever took the initiative to redescribe in the opposite direction did not realize the problem that was created by making the variable course in the closing portion of the first description a positive call when used in the opposite direction. The lesson here is that one should not reverse the direction in which a description has been written. This does not mean, however, that you cannot plot the description in reverse in *looking* for a solution to a problem in an exceptional situation.

Excess or Shortage

Another point along the same line is that you cannot take an excess or shortage found in a description and prorate it among all the courses described. There are several reasons for this. The first is that when a course is described without a tie, or the limitation of an adjoiner, then the amount of the cited bearing and distance must hold. Second, if you have an excess or shortage in the total traverse, then you must search through each course, reviewing and comparing field work and record information to find and isolate any mistakes resulting from omissions, transpositions, errors, etc. Third, if you have an excess or shortage in the closing course, then it belongs there until a field measurement fixes the correct amount. Fourth, it is the generally accepted rule that proration is to be used only as a last resort and only when sufficient correlative information is found to make it a plausible solution.

When a map of a survey has been submitted to you and this information shows that the distances to controls cited in the deed have been measured longer or shorter than the amount in the previous description, this information should be incorporated in the re-writing of the description at hand, thus reflecting the facts in the field. Make sure also that you do not change those items which are already on record and if they do not make ties to any particular monuments or adjoining deeds, do not introduce new ones and claim them to be of record. On the other hand, analyze the maps and determine if the surveyor tied to occupation or anything else which was not called for in the description on record. Always go back to that which the owner acquired by his deed.

When lots or parcels are created simultaneously by the filing or recording of a map, problems sometimes arise subsequently as a result of measurements not in accord with those shown on the map. This is more prevalent in the case of older maps which were made from less carefully done field work and the greater accuracy of today's work reflects the difference. This condition also occasionally develops when a group of metes and bounds descriptions are recorded simultaneously. Theories for the adjustment of excess or shortage can be found in any number of books pertaining to survey work involved in the reestablishment of such corners and lines on the gound.

However, even though the proportioning of an overall distance amongst its parts is an easy way to dispose of the excess or shortage, such a method is to be employed only as a last resort. This approach is further confirmed in the action of the courts by the fact that there are relatively few cases which have actually concluded their decrees with the direct use of the proportionate rule. The exception to this is in the occasional application to the matter of small differences within a limited area between acceptable found monuments, or, in following those certain rules governing the Restoration of Lost or Obliterated Corners and the Subdivision of Sections as set forth by the U. S. Department of Interior, Bureau of Land Management.

Apportionment Theory

These problems are generally categorized as belonging to the apportionment theory or the remnant theory. The first in the absence of monuments is the application of proration of the total measured distance of a block of lots to the total record distance into the aliquot parts, or, if proper monuments are found at some lot corners and not others, the measurements between those that are further apart than one lot width may be prorated against the record. This is used with both straight and curved lines.

Remnant Theory

The second theory concerns the odd lot which is usually at one end of a block, although in a rare situation it may be in some other location because it fills the odd shaped void next to a diagonal or curved street or boundary left after all the regular shaped lots have been put in place. Because of this "left over" status, the lot is sometimes shown on the map with no dimensions and in Pereles v. Gross, 126 Wis. 122, 105 NW 217, it was held that it is presumed that the surveyor placed whatever the remainder, "*much or little,*" into that irregular undimensioned parcel. If the remnant lot is dimensional on the map, its positional determination must consider the occupational status of the other lots before a final decision of its size. In this regard, there have been court cases decided both on the apportionment theory and on the remnant theory.

Insofar as writing descriptions is involved with these two theories, the simple reference to the lot shown on the map is still sufficient as far as title for the whole lot is concerned. The problem will arise in case of a partition or a request for a perimeter description of any lots subject to these theories. If you are asked to write a perimeter description, include a tie to each corner of the lot by reference to the map and its recordation, for safety and preservation of title.

If you are asked to write a split decription of a remnant lot, study the overall picture first and see if you are satisfied with the facts presented; an inspection of the physical conditions might be in order to determine their intercompatibility. And again, look to any court cases in your area to determine if a precedent of this nature has been established. A single line type of description would probably be preferable, assigning "that portion of lot___, etc., lying east of the following described line:" for one part and "that portion of lot___, etc., lying west of the following described line:" for the other part using survey ties to an outside control or a boundary street.

In a different vein, there are a number of cases that have dealt with land not acquired by a subdivider. Without considering the reasons why it was done, the fact is that the map prepared by the surveyor or engineer included more land than was acquired by the description in the deed to the developer. Notwithstanding the length of time that subdivision has been officially on record, the subdivider has no right or title in the excess land included on his map or on the ground over and beyond his lines of title, and therefore does not possess the right to convey said excess land.

This matter has come before the courts several times and they generally concur in the fact that title to those parts of the lots shown on the tract map that are outside of the boundary acquired by the developer did not pass and the buyers had recourse to remedies for misrepresentation. Perfection of title in and to those lots would be accomplished by acquisition of the extra land by the subdivider from the adjoining owner, or in case a title company has insured the subdivider, it would probably assist in the acquisition.

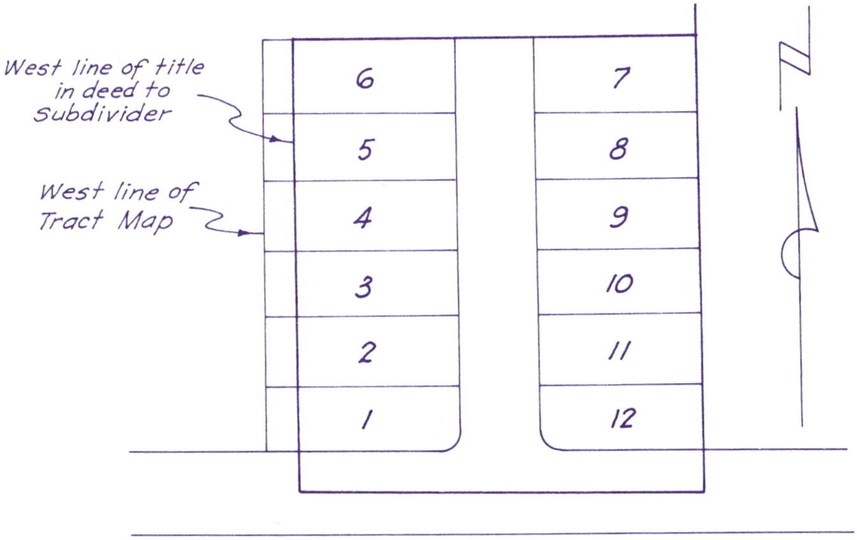

Fig. 47

In Figure 47, title to the westerly portions of Lots 1-6 lying west of the deed line does not pass to the buyers of those lots from the subdivider.

The discussion on excess or shortage has thus far centered on such conditions relating to or controlled by distances but areas are also involved though not as often. If, for instance, the quantity of each parcel is shown on a map, recorded or not, and the distances do not check either on the ground or by a mathematical non-closure, the area(s) will be brought into play for correlative support of a correct answer in distance(s) and perhaps bearing(s) too.

In other words if all the lots on a map were designated as "10 acres" each but one set of facts produced less acreage while another method resulted in being very close to 10 acres, the latter would be presumed to be the proper solution because it more nearly satisfies the original plan. With all lots shown on a map given various acreages, the application of that measurement adjustment which would result in acreages respectively proportionate to those on the map would be confirmation of a proper method. This applies to both public land and private land surveys.

In the situation of non-conformity between adjoining properties described by metes and bounds which include an assignment of quantity in their deeds, there may also be in case of dispute, the application of the proportionate theory of area in any effort to determine the correct solution regarding the questionable measurements.

Net or Gross of Street Area

Questions arise when the wording is not sufficiently explicit. Sometimes the reference to the lot corner will be made without qualifications, but when the description is read out on paper and boundary descriptions are run into the picture, the decision must be made as to whether the lot corner referred to was in fact in the center of the street or on the side line.

This type of situation calls for clarification and if it is known and intended that the description be developed from the centerline of the street instead of the side line of the street, then it should be clearly stated so.

A street and a lot are two separate and distinct terms.[17] When you talk about a lot, you are relating it as bounded by a street, alley, walkway or whatever public area is next to it. So reference to "that portion of Lot 3 described as follows:" does not include any part of the street or other public way adjoining it insofar as the description is

[17]Montgomery v. Hines, 134 Ind. 221, 225 (33 N.E. 1100).

concerned. In other words, this descriptive format is not concerned with the part in the street which the owner of the lot may or may not have. This will be discussed in more detail in Chapter 9.

It has been held, however, that if it was the custom in a certain area to buy or sell property on the gross basis (described and measured to and from the center line of the street), then the custom should prevail. Contrariwise, if it was the custom to sell on the net basis (side line of the street), then that custom should prevail in that area.[18]

The essence of this whole matter is that the wording in the description should be so clear that there is no question as to whether the measurements are to be made from street side line or center line.

If the map of the subdivision of the land shows that the solid side lines of the lots stop at the side lines of the streets, then it is clear that the lots are distinct from the streets. There are some maps, especially older ones, however, which are ambiguous in that they show the side lines of the lots going out to the center lines of the streets and furthermore, all lines are solid lines. This might indicate that the lot includes the street but it is not certain. Some maps carry a statement, "The areas and distances shown on this map go to the center line of the street." It is very helpful to know in fact what the engineer or surveyor did but this statement alone does not move the lot line from the side line of the street out to the center.

Occasionally a map will show the side lines of the street as dash lines across the lots. This indicates an easement type of street, or a private street, rather than one with the full effect of dedication.

If it is desired to have the description carry to the center line of the street, it must clearly state so in some such manner as follows:

> "Beginning at a point on the west line of said lot being, for the purpose of this description, the center line of 'A' Street adjoining said lot; thence . . . ." or
>
> "Beginning at the northwest corner of said lot, being the center-line intersection of 'A' Street and 'B' Street for the purpose of this description; thence . . . ."

Obversely, if the map is indefinite or there is any question of meaning and the side line is to be held, rather than the center line, use a reverse form such as:

> "Beginning at a point on the west line of said lot being, for the purpose of this description, the east line of 'A' Street 60 feet wide as shown on said map adjoining said lot; thence . . . ."

[18]Ferris v. Emmons, (1931) 214 C 501, 6 P 2d. 950.

This will retain the *original* street line established by the map regardless of how many widenings may have been taken because sometimes the widening is greater on one side than the other.

Indeterminate Calls

It is unfortunate that some descriptions require so much time to be wasted in analyzing the meaning of indeterminate calls. Some are easily picked out such as, "westerly parallel with the westerly line," which is of course impossible. However, when reading, "along the easterly line of said lot to a point 203.23 feet to a point 12.45 feet from the corner of Jones' land," you wonder what they are attempting to do with no directions given. Also, when you find ". . . . and running thence south 2° 29′ 20″ East along the easterly line of said lot 16, 242.58 feet . . . .," does this mean 16, 242.58 feet along the easterly line of the lot, or 242.58 feet along the easterly line of Lot 16? This is a good illustration of the reason why you should *not* put distances next to a lot or block number. A good rule to follow is to keep the distance immediately next to the direction. In this way, you eliminate problems in the communication of correct facts. This not only applies to straight lines using a bearing and a distance but also to curves using a direction and distance. Let any qualifications applicable to a line or curve either precede or follow in toto the combined direction and distance.

Majority Probability

For correct title interpretation, assemble all the facts including record and off-record information and occupation as determined by survey. From this data, set forth all the probable and possible theories of line and corner positions giving full consideration to the physical, engineering, title and legal factors and their interrelationship to ascertain a solution which will fit the majority of the analysis most closely. This is the theory of *majority probability,* and normally will produce the most satisfactory results.

The Meaning of "One Half"

Many times owners wish to divide their property in half or buyers wish to buy a certain one-half of what the owner has. In either case, the description must be correctly worded in order to pass title to that which has been requested.

The rules for constructing a correct description for this purpose (and the method of determining its location on the ground) depend on the type of land; namely, whether it is public land or private land.

1. Public lands are those acquired by the Federal Government such as the Louisiana Purchase from France and the Mexican Territory by the Treaty of Guadalupe Hidalgo, except for certain parcels which had been previously granted to individuals. Subsequently, the public lands (not included within private grants) were surveyed and subdivided into sections according to certain rules formulated and set forth by the Federal Government's Bureau of Land Management. Those rules prevail in the subdivision of the land for sale and for the reestablishment of lines and corners.
2. Private lands are those not held by the Federal Government and not subject to its rules of subdivison, examples of which are ranchos or any part thereof, private tracts granted prior to subdivisions of section land, parts granted from the government into private ownership after its land has been sectionized, lots in a private subdivision, colonial tracts, etc.

Rules for Dividing Land into Halves

1. Public (government) lands follow the rules set forth in the "MANUAL OF INSTRUCTION FOR THE SURVEY OF THE PUBLIC LANDS OF THE UNITED STATES, DEPARTMENT OF THE INTERIOR, BUREAU OF LAND MANAGEMENT."
2. By court rules governing private lands, each half must contain exactly one half of the total area.

Application

1. Public Lands — Government Rules.
   a. If monuments exist on opposite sides of the section in the area midway between the outside corners, a straight line is drawn between the found points, wherever they are. See Figure 48. This may or may not divide the section into two equal areas.

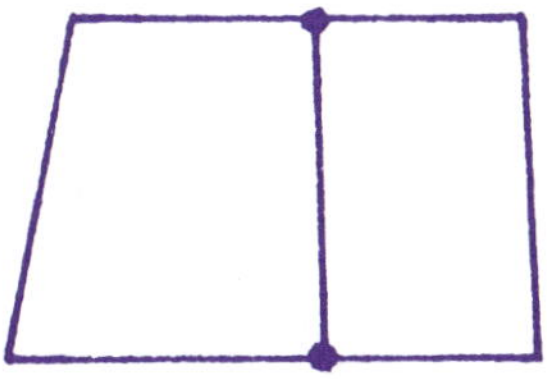

Fig. 48

   b. If it is a regular section with no existing monuments, then a point is set at the exact mid-point on each opposite side and a straight line is drawn between them. See Figure 49. This also may or may not result in two equal areas.

Fig. 49

c. If it is an irregular section (usually called a fractional section) with no existing monuments at the so-called mid-points, then the midway points on opposite sides will be reestablished by proportion to the original survey and a line drawn between those two points. See Figure 50.

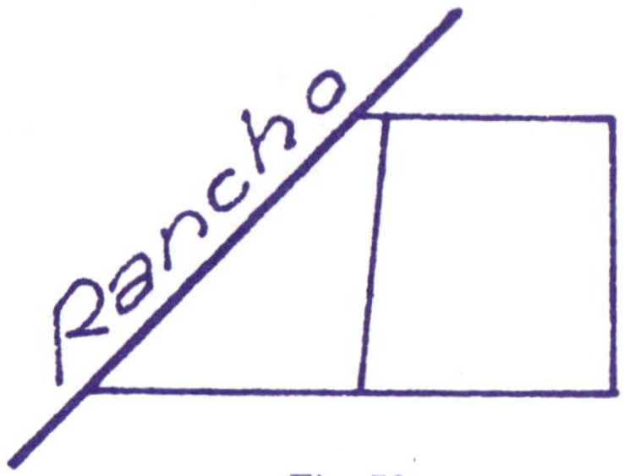

Fig. 50

2. Private lands — there are *no* rules establishing the *position* of the line between halves if the dividing line is not delineated.
   a. If the description is, "E 1/2 of Lot . . . ." with no delineation of position of the line between the halves, the position of the line is indeterminate; it might be parallel with one line, right angles to another or even curved so long as each part has exactly one half the total area. One man made an arbitrary division of a lot into halves and made it hold as illustrated in Figure 51.

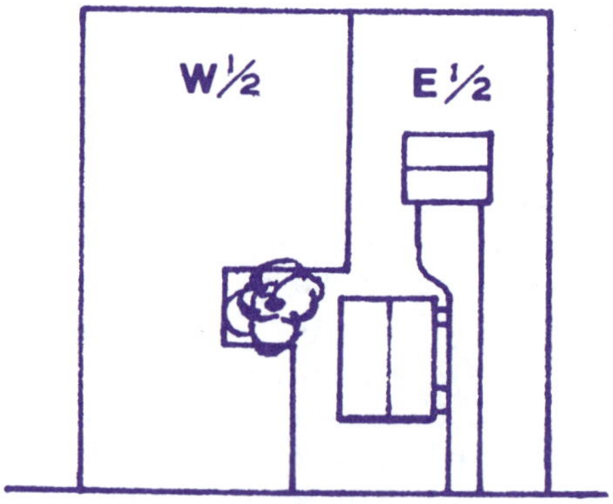

Fig. 51

An interesting court case in San Francisco, California, concerned this principle. A man had a large lot capable of supporting two home sites within the regulations of the zoning ordinance. He wanted to go to Europe so he decided to sell one half of his lot to finance his trip. He sold the east half, took the money, and enjoyed his visit. When he returned, however, he found that the man who bought the east half had built a home and put a fence along the west side of it which was not on a straight line, and neither was it in the middle! So, he asked him, "How come your fence is not on a straight line?" The man said, "You sold me the east half of your lot, but you didn't say where the line was between the two halves and I have not taken any more than my half." The seller said, "Well you can't do that. I sold you right down the middle on a straight line from front to back." The man said, "No, you didn't say where the line was; you didn't put that in the deed." The seller said. "Well, we'll see about that." so he went to court.

Naturally, the court required a survey and the survey substantiated the buyer's claim; namely, that each side had exactly one-half of the lot area. In putting up his fence, the buyer wanted to include a tree; so he went part way back, then jogged around the tree, and then cut back further on the rear of his part of the lot and left more in that area in the back to seller on the west part. The court dismissed the case on the grounds that nothing in the deed of conveyance nor the escrow stated where the half line should be and the buyer had not usurped any more than his half; therefore, he was within the law. From this you can see that it is essential to have the line between halves delineated.

3. Private Ownership of Government Lands
   a. If one person acquired all the Northwest Quarter of a fractional section, he could sell the North Half and the South Half on an even split of area basis if he wished to as a private owner, or, he could sell the parts on the same basis as the showing on the Township Plat. See Figure 52. This was upheld in a court case which said the individual owner of a whole part was free to choose his method of subdivision.

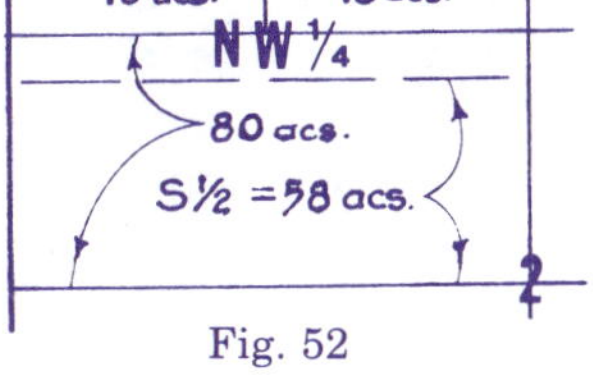

Fig. 52

b. Another example of upholding the one-half by area rule for privately owned land is found in the following case:[19]
In Section 30, T 20 S, R 24 E, M.D.M., Tulare County, California, there is 178.98 acres in Southwest one-quarter per the original government survey. A subsequent survey showed this quarter as containing 180 acres more or less. The entire Southwest one-quarter was patented under 1 entry as 1 tract in 1872. On 10/22/09 the owner of this tract (Plaintiff) conveyed:
"The East one-half of the Southwest one-quarter Sec. 30, etc. . . ." with no acreage recited. The buyer (Defendant) took possession of 90 acres. See Figure 53. Action was filed to get possession of the extra 10 acres not included in the east 80 acres as it was shown on the Official Township Plat. The trial court said 1/2 = 1/2 of 180 acres. The appellate court confirmed the decision so the buyer retained the 90 acres.

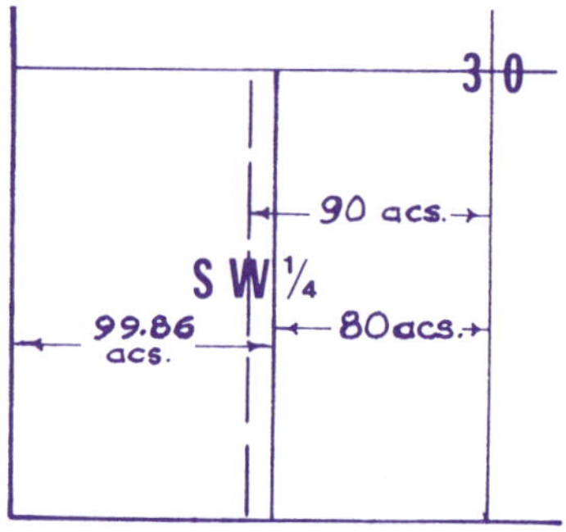

Fig. 53

c. On the basis of a misconceived assumption, the owner of the southeast quarter of section 10 sold simultaneously two parts thusly:
The west 80 acs of the SE 1/4 Section 10. . . . .
The east 1/2 of the SE 1/4 Section 10. . . . .
The seller thought he had divided his land equally but there existed an excess in the quarter section so a gap was left between the two parcels. Unfortunately, these facts were not discovered until after the grantor was deceased, thus requiring legal determination for the dispensing of the excess.

[19]Wood v. Mandrilla, 167 Cal. 607.

In the final analysis of descriptions, you must carefully consider *all* facets of meanings and interpretations before rendering a decision.

Unwritten Words

Although the courts have held that parol and/or extrinsic evidence, in other words, evidence *aliunde,* is admissible to assist in determining the intention of the parties to a document, there are instances where the lack of certain words does not detract from the meaning of the description.

Right Angles

Going back to the assertion that as between individual parties, the grantee is favored, consider the connotation of a given distance between two parallel lines; that measurement is sometimes qualified with the statement, "measured at right angles," to be sure that the space between the two lines is at the maximum possible distance. Because of court decisions to this point, it is unnecessary to include that qualifying statement because the control will still hold.

Tangent

In regards to curves, the word "tangent" or "tangential" is often used in various juxtapositions such as "to the beginning of a tangent curve," "to the end of said curve; thence along a tangent line," "along a tangential curve," "thence tangent to said curve," etc. The fact of the matter is that a line and succeeding curve, a curve and succeeding line, compound curves and reverse curves are each tangent to each other *without saying so.* It is in the case of *non*-tangency that a specific statement should be made to clarify the situation.

When a description reads, "to the beginning of a curve concave to the north and having a radius of 200 feet; thence easterly 52.64 feet along said curve; thence N 86° 22′ 30″ E 457.38 feet;" it is understood without words to the contrary that the curve is tangent to the preceding line and the following straight line is tangent to the curve; therefore, the *absence* of the word "tangent" does not defeat the meaning.

Prolongation

Another case in point is relative to parallel lines where the statement is made, "northerly to a line parallel with and 25 feet southerly from the easterly prolongation of the northerly line of Lot 5." This could be written without "the easterly prolongation of" and the absence thereof would not defeat the meaning.

"NAP"

On the maps of some subdivisions, one or more areas may be included within the lines of the tract but not established as, or given numbers as, lots in the tract. They are labeled "NAP" or "NOT A PART" or "Not A Part of this Subdivision." For one example, this occurs when the seller excepts his homesite from the conveyance of the large parcel to be subdivided. In the layout of the new tract with lots set around three sides of the homesite and a street in front, all the boundary measurements of the "NAP" are shown on the tract map making it eligible for conveyance by direct reference in a document. The form of description to use for the "NOT A PART" area will be explained in Chapter 11.

Intention

After finding such items in land descriptions as insufficiency, ambiguity, conflicting elements, mistakes, indeterminate calls, etc., one should not overlook that certain avenue of qualitative analysis that seeks to solve the questionable matters by learning the *intent* of the parties involved. This is not just a simple matter of asking them because, in many instances, the originators of the transaction are long gone by the time the problem rises to the surface due to its financial value.

Obversely, however, when all the facts in a description are finite and conformable, you have control properly and well expressed so one does not need to look for some hidden meaning as an intention. One person I recall tried to create a problem out of the call, "N 23° 25′ E. 256.6 feet to the line of the fence set 29.6 feet southwest of Hokum's barn." by claiming that "while the distance to the fence measured 252.0 feet, the distance of '256.6 feet' should hold because it would leave 25 feet to Hokum's barn and that is a normal amount and is all he really needs therefore that was probably the intent." Since the citation of a physical monument as a tie in the legal description carries a number one priority in value, this would be a case where there is no need to even consider the alternative of intent.

From this a conclusion can be drawn that only when the wording is *questionable* does one look for intention. Judge Savage of Maine further pinpointed this quality as ". . . . the intention effectually expressed, not merely surmised."

Intention developed from construction of the wording in the deed:-it is a correct presumption that the creator(s) of the deed (there may have been more than one working at it) intended it to be valid. From that basic concept, one must analyze the meaning of the words actually used in the description in such a way as to result in legal effectiveness instead of a void.

It is a general application of the rule that course and distance give way to an adjoiner. Whether it be a marked line as shown on a map and staked on the ground or an unmarked line described in a recorded deed, the called-for line is considered as a monument and therefore more certain than the course and distance; however, the contrary may be shown. If it can be determined that the tie to the adjoiner was made by conjecture, or erroneously, it must give way to the course and distance if that complies more closely with the intention, especially if the latter is locative and certain.

It is a long-established adage that that construction of the wording in a description which renders the most certain conclusion is the one to be adopted. Contrariwise, sufficient ambiguity will make the deed void. In between these two situations is a matter of degree. In every case, the description must be read and considered *as a whole*.

Interpretation of individual (including corporate) deeds will show favor to the grantor. On the other hand, intent concerning the meaning of a description in a will has been held in favor of the testator because he had certain plans to be carried out.

Construing Descriptions

Although the practice of surveying relies in part on the application of mathematical formulas, rules of construction, etc., the professionalism of surveying is manifest in the discretionary use of such knowledge, coupled with experience.

When a parcel boundary or a strip of land is described with sufficient detail and with proper ties, there is no problem. Sometimes, however, in an effort to make sure that all possible ties and references are included, the result is conflict because some of the information is incompatible with other parts.

A good description writer will check all references *before* using them in order to assure compatibility.

Where two descriptions purport to cover the same land but between which there is repugnancy, the court will review surrounding facts and favor the most definite and those which confirm the evident intent. Certainly where one is ambiguous, the other more certain will prevail.

If reference is made to a map which is inconsistent with other particulars, the map will be given preference if the parties to the transaction accepted the map as their common understanding; otherwise, that map is considered as subordinate to ascertainable and definite facts.

PROBLEMS

1. Could insufficiency in the description cause a problem in the perfection of title?
2. If an adjoining deed holds priority but there is no mention of it or tie to it in the instant description, will that boundary of the adjoiner prevail over the course and distance recited without it? Give the reason for your answer.
3. Is a map or a plat considered to be a monument?
4. Do you consider that field notes would prevail over a map prepared from them in the case of dispute as to interpreting a description? Explain.
5. What value does the reference in a description to a "surveyed" line possess as compared with a reference to a monument?
6. If any part of a lot in a subdivision lies outside the lines of the description in the deed to the subdivision, what is the disposition of title to the buyer of that lot?
7. What are the two theories used relative to the adjustment of lot sizes when ground measurements differ materially from those shown on the map?
8. When reference is made to a part of a lot, does it include a part of the street?
9. In a metes and bounds description, is it advisable to place the distance immediately following a lot number?
10. What are the two categories of land which control the method of dividing property in halves?
11. What is the one rule for dividing "private" land into halves?

Chapter 8

Monuments

This word brings forth different pictures to different people. To some it means a statue or memorial to a great person or a renown event. To others it brings to mind a stone, a tree, a post or a fence. The surveyor remembers these also, but, in addition, he thinks of stakes, ironpipe, brass caps, bronze plates, nail and tin, lead and tack, PK nails, spikes and chiselled crosses in rock, cement, etc. On the other hand, attorneys consider not only physical monuments but those that carry legal connotation and are used in descriptions and by surveyors for the control of boundaries.

Monuments of whatever nature assist in holding together the framework of property delineation, both on paper and on the ground. The discussion here will consider not only the value of various monuments but their relationship in juxtaposition to legal descriptions, occupation and surveys.

The question may be asked as to why a scrivener should have an interest in monuments; he neither finds them nor sets them. The answer is that one who interprets descriptions of land must know first the relative value, perhaps better classified as quality, of any pertinent monument and, second, from this, its applicable authority to control, or not control, a line or corner.

A monument is a "thing" recognized and given authority for the purpose of indicating or referencing a point or boundary to be used for control. "In real property law and surveying, monuments are visible marks or indications left on natural or other objects indicating the lines and boundaries of a survey." Black's L. D. "Permanent landmarks are established for the purpose of indicating boundaries.": Bouvier L. D.

Examples

Monuments are classified in two general categories — legal or physical. Legal has to do with those matters which are in references deposited in the form of documents in public offices. Physical refers to both natural and artificial physical things which are accepted and used as marking boundary lines and corners.

Documentary papers such as deeds for a railroad or canal right of way or easement deeds for a highway, pipeline, power line, etc., are considered as legal monuments. Reference to documents of ownership such as a tie.

"to the east line of the land described in deed to Smith. . . ."

is considered a call to a record monument and being a matter of public record makes it a legal type of monument. Often the reference is shortened and reads merely, "to Smith's land. . . ." This brevity may be indicative and sufficient enough unto those who are immediately concerned at the time of the transaction (as explained with examples in Chapter 3) but reference to the complete detailed information; viz, "The land described in deed to Albert Smith recorded January 16, 1918 in Book 315, Page 42 of Official Records" is far more satisfactory. This prevents ambiguity in case Smith owns more than one parcel of land, and furthermore, the deed by this reference can then be easily found in the public office of records.

The courts have held that the line of a described adjoiner "is a fixed natural object."[20] and "a call for the line of another tract of land is a natural boundary." However, having established the adjoiner as a boundary does not in itself determine all of the ramifications of that

[20]Fidelity Realty Co. v. Flahaven Land Co., 193 Ky. 355, 236 S.W. 260.

boundary.[21] The description in the referenced deed might be defective.

A tie to a road described in a deed or shown on a map, either to the center line or side line, is also a tie to a record monument. Any such call, however, should definitely state to which line the tie is made so that there is no question as to the intention of whether the total description is a net or gross area as mentioned in Chapter 7.

Also, the question as to whether the road strip as a monument carries the status of fee or easement is another matter. When it is created through the medium of a grant deed without qualification and carries no addendum clause, its condition is a fee absolute. If the strip is granted on an "EASEMENT DEED" or on a regular deed form which recites "A right of way for . . . ." then it is clearly an easement. In between these two conditions, there are some documents on record which start like a grant of fee and end with a qualifying clause (or clauses) as an afterthought which results in making the status of the conveyance questionable.

Generally when a street, or any other public way such as a walk, alley, etc., is created on a map and thereby dedicated to public use, it carries the status of a non-exclusive easement with an underlying interest usually divided between the lot ownerships adjoining each side of the street. However, there are some areas where a map duly recorded in the public office carries the effect of transferring fee to the political entity for those portions designated for public use.[22] This means that you should check the code of your state and local ordinances to know which rules prevail.

The scrivener needs to comprehend these facets of meaning in order to know whether, by the description, title in the street[23] is included, excepted or questionable.

Court Cases

Although the general assumption supported by various court cases is that the underlying fee title normally carries to the center of an adjoining road easement, similar to the idea of ownership to the thread of a stream, attendant conditions, information shown on the map and the format of the qualifying words are in the final analysis the determination of the status of title attached thereto.

[21]Cherry v. Slade, 7 N.C. 82, Graybeal v. Powers, 76 N.C. 66.
[22]Sears v. Chicago, 247 Ill. 204, 93 N.E. 158.
[23]"Road," "street" and "highway" are considered as synonymous for discussion here.

A street, with or without adjoining lots, created by a subdivision of land which is shown on a map is a legal monument. A description of a part of the lot; viz, "the southerly 50 feet of Lot 2" carries with it the control of 50 feet measured from the north line of the street adjoining. This is on the basis that the deed refers to that portion of the lot set apart for private use and not any part of the title interest underlying the street for public use.[24] In the case of Earl v. Dutour (1919), 181 Cal. 58, 183 P 438, the street and lot were considered as two distinct and separate terms; this in view of the fact that the owner of any such lot has no right to the possession or occupancy of any portion of such public street.[25]

In the case of Ferris v. Emmons (1931), 214 Cal. 501, 6 P2d. 950, a substantial amount of parol evidence was introduced, "both as to common reputation and custom in the community of Pomona. . . . . that it was commonly understood that 'the blocks extend to the center of the streets.' " The court found and held that if the understanding and the custom of transferring property in a certain area was based on the center of the street, then that custom should prevail.

One way to check such a condition for an existing custom would be to take all the deeds within certain bounds and draw them according to their descriptions. If they fit all together on a net (side line of street) basis, then that apparently is the custom and if the obverse is true, then the gross basis prevails. Remember, however, this matter of net or gross consideration is relative to the sale of the land and does not change the hereinbefore cited distinction between the lot and the street.

Any number of cases have settled the fact that streets around a block of lots hold their position as physically developed and excess or shortage is allotted to the holdings between them. Whether it be prorated throughout or applied in entirety to one part or another depends on the circumstances involved.

Sometimes a change of controls is made inadvertently, while at other times it is done purposely. Consider a certain block where a survey of an interior lot revealed the fact that, although the overall measurement of the block between the streets as monumented by the city approximated the showing on the record tract map, the positioning of the lots relative to the streets as held by the city overlapped

[24]Peake v. Azusa Valley Savings Bank (1940), 37 CA 2d. 296, 99 P2d. 382, refers also to Wegge v. Madler 129 Wis. 412, 109 NW 223.

[25]Wegge v. Madler, 129 Wis. 412 (109 N.W. 223).

driveways and buildings. A search from lot to lot uncovered a majority of the 2 x 2's set when the tract was created, and their measurements showed a picture of conformity within the tract except for the lots on each end of that block because the streets at each end had been improved and their center lines monumented on a shift of several feet longitudinally. This was a case where the 2x 2's had sufficient weight to hold over and beyond the normal control of the streets.

This indicates the right and prerogative of the engineer or surveyor of a political entity to establish and maintain street improvements on an alignment other than the center line of a right of way shown on a map, or different from a section line, whether it is for the objective of having a straight line or for some other reason, but remember that any such realignment does not of itself change or move any lines of title.

Another example, however, must be relegated to accidental location because the road department took the wrong stakes and laid the pavement some 36 feet from the quarter-section line which had been designated for the center line of the road in the deed. The center line of this physical roadway, then, obviously could have no value as a legal monument control for determining property lines. It is another of the many cases where one line on the ground may be used for improvements, but another line must be used for the title control of any surveys of land adjoining that street.

One case in particular cites, "there is a presumption that all those thoroughfares are where the dedicator and the municipality fixed them and dedicator's privies are bound as he would be."[26]

If, on the other hand, a road is created by a document calling for "an easement over the west 30 feet," then any description that is intended to stop at the side line must clearly state the fact. This can be accomplished either by tieing to the side of the road or by going to the center line and then, "EXCEPT the west 30 feet." Connotations of different meanings resulting from different combinations of words concerning title in or out of a street are given in Chapter 9.

Natural physical monuments are such as a river, tree, lake, pond, cliff, mountain ridge, ocean shore, beach, etc. Even streets and highways have been classified as "natural" monuments by many court cases where the physical condition has been "opened and long acquiesced in." The call for the center of an extension of a street is not eo nomine a fixed monument, where the street up to that time had not

[26]Cardosi v. Wise, et al., 44 Ohio App. 206.

existed,[27] but an established avenue is a monument which may properly be used to mark boundaries.[28]

Other items which have also been classified as natural monuments include such things as "a clearing," "bottom," "spring brook," "creek," "wash," "gully," "gulch," "burying hill," "culturable lands," "swamp," "the run of a swamp," etc. Because of their more permanent nature, all of these are usually considered superior over artificial monuments except that in Perry v. Pratt, 31 Conn. 433, it was held that "A tree may be torn up by the roots in a storm, an imbedded stone dug up and removed, a stump be obliterated by decay, and the course of a stream be changed by avulsion, and when that which apparently was fixed and stable becomes displaced and unstable, it may no longer be imployed in conveyances to define boundary lines."

Sometimes there is a question of priority concerning two similar conditions. It has been held that where an expression equally fits two natural objects, one a matter of record and the other not, that of record shall be understood as intended.[29]

This same idea applies to maps in the light of record monuments; in the case of two maps, one recorded and the other not, the one on record will receive preferential acceptance. In view of this, the surveyor who files his map as a public record (assuming that it bears sufficient evidence to support his position) will carry a preponderance of weight in any matter of dispute. This is also why references in descriptions should be made preferentially to matters of record.

Artificial monuments are "things" cut, created or formed by the hands of man such as stones, wood posts, iron pipes, concrete posts, rods, stakes, etc. In contrast with references to natural monuments as they exist, artificial monuments are put in a particular place to represent a definite point to be used as a boundary line, an actual corner or a witness point to a fact. The law for surveying public lands and the value of monuments set therefor with all the ramifications thereto is set forth in the MANUAL OF INSTRUCTIONS FOR THE SURVEY OF THE PUBLIC LANDS OF THE UNITED STATES prepared by the BUREAU OF LAND MANAGEMENT, UNITED STATES DEPARTMENT OF INTERIOR. In this Manual (some times referred to as the "blue book"), it states that

"The law provides that the original corners established during

[27]Guitar v. St. Clair, 238 Mo. 617, 142 SW 291.
[28]Kidder v. Smith (Mo. A.), 180 SW 1061, 1062.
[29]Mosby v. Carland, 4 Ky. (1 Bibb) 84.

the process of the survey shall forever remain fixed in position, even disregarding technical errors which may have passed undetected before acceptance of the survey."

The courts have upheld the maintenance and use of those monuments in finality as set by and for public land surveys. This might be the reason why it is sometimes assumed that *every* monument is good and conclusive. It should be clearly understood at this point however, that outside of public land surveys, monuments set or found by private surveyors[30] are not necessarily always acceptable.

In Chapter 4 it was stated that monuments have priority over distances and bearings, but this is held as a part of the rules of construction used to ascertain the intention of the parties.[31] When, as or if monuments are placed or referred to by mistake and it becomes apparent by analysis, then the rule that artificial monuments control over distances and bearings does not hold,[32] nor will they hold in opposition to intent.[33]

Value

Before one approves of the use of a certain monument, either for survey location or control in a description, it should be subjected to an analysis on the same basis as those individuals who want to get into the elite "400" Club. Each one must have background, character and recognition before being accepted. A simple illustration of this is a section corner stone; if it has been described in the surveyor's field notes, it has background; if it carries the markings called for, it has character; if, in addition to these, it conforms to position in relation to other objects, it may then have recognition and by the sum total of this analysis, it can be accepted. Any questionable monument should be subjected to such a review before accepting it.

The doctrine that coterminus owners may establish their mutual boundary line by agreement, which includes implied acquiescence, is applicable only where the true line is unknown or uncertain.[34]

"A boundary is considered definite and certain when by survey it

[30]Private surveyor in this context includes official surveyors of the city, town, county, parish, etc., in contrast with the public surveyor of the federal government.

[31]Green v. Horn, 207 N.Y. 489, 101 NE 430.

[32]Ramseaur v. Ball, 59 Tex. Civ. A. 285, 125 SW 590, Ralston v. McClurg, 9 Dana (Ky.) 338.

[33]Townsend v. Hayt, 51 N.Y. 656, Nissley v. Moeslein, 23 Pa. Super. 119, Murdock v. Chapman, 9 Gray (Mass.) 156, 75 Mass. 156.

[34]Establishment of boundary line by acquiescence, notes 113 ALR 421, also McK Dig. Ref. Bdrs. Sec. 31.

can be made certain from the deed,"[35] In this respect, there are examples of "created" monuments which have fallen by the wayside, both with and without court cases to trip the prop from under the point. By "created" monuments is meant the setting of new physical monuments to make a line newly described in relation to a greater control line, such as setting the west line of the east half of the southwest quarter of a section for the first sale out of the said southwest quarter, or, such as setting the lines of lots in a newly-created subdivision.

Under this consideration when a common grantor creates a new line and a physical location is made, the recipients or grantees are bound thereby,[36] and this line is presumed to be the one which the words in the deed described.[37] However, this very creation becomes one of the most common problems of today's engineer and surveyor in establishing a relocation because, in searching for points upon which to develop the framework of the survey, every old pipe, stone, stake, bolt, nail or what-have-you stands out, and each offers itself as a ready hook upon which to tie the survey lines, but the discernment of the monument is the question.

In the eyes of the law, chronology plays an important role. If monuments can be proven to have been set immediately prior to or coincident with the creation of the description, they are held to be emblematic of, and the physical control of, the written words, and more especially if cited. In contrast, if the timetable shows a material spread between the creation of the deed in words and the subsequent setting of the points on the ground, the monuments cannot be claimed as superior to the calls. Monuments must be fixed and certain and their position at the time of the conveyance must be determined in order to pinpoint the location of the description. An extreme condition is cited in Payne v. English, 79 C 540, 21 P 952: "The line of a street covered by the waters of a bay, and having no definite dimension or boundary at the time of the conveyance, cannot control the other descriptive parts of the conveyance of a water lot."

The analysis of the worth of a particular monument can be developed from several standpoints. One of the primary considerations is the question of originality; is that monument found one which was

[35]Janke v. McMahon, 21 Cal. Ap.. 781, as cited in Williams v. Barnett, 135 C.A. 2d 607, 287 P. 2d 789.
[36]Turner v. Creech, 58 Wash., 439, 108 Pac. 1084.
[37]Herse v. Questa, 100 App. Div. 59, 91 NYS 778.

set by direction of the common grantor? If it is, then it can be presumed to be representative of the words in the document of transfer, and vice versa — unless sufficient and overwhelming evidence is available to prove the point to have been set inadvertently in error. Next, if the point is not an original, is it a replacement — that is, a superimposition and not a subsequent relocation? If it is a replacement and proper records verify it, then the same value may be attached thereto as for the original. If, however, the point is a relocation instead of a replacement, careful analysis is in order, for determining its value relative to the original, before accepting it. "If monuments once existing are gone, and the place where they originally stood cannot be ascertained, the courses and distances, when explicit, must govern . . . . "[38] This is taken to mean that unless a precise location can be made, definite courses and distances are to be preferred.

In making use of monuments, remember that a prime requisite for a monument to control over course and distance is to have it recited in the deed ". . . . courses and distances must yield to monuments, visible objects, or natural boundaries called for in the deed."[39] And conversely, "while it is true that where courses and distances mentioned in the deed conflict with fixed monuments therein mentioned the referred to monument will control, the rule has no application where surveys, stakes, or other monuments are not referred to in the deed."[40] In other words, if a course and distance go "to a 4″ x 4″ pinewood post on the brow of the hill," the monument holds; but if that course and distance does not call to any definite thing and there happens to be a 4″ x 4″ pinewood post within a few feet of the end of that course, the post will not necessarily hold. If there is any doubt as to the location of the monument, the reason why it should control ceases.[41] In some few deeds, each course goes "to a 2″ iron pipe" or some other type of monument which gives the relocator some particular thing upon which to rely. It is an especially good way to write a description where an agreement of sale has been based upon a positioning and marking of the corners on the ground by mutual understanding.

Visible and ascertained objects as monuments govern over courses

[38]Drew v. Swift, 46 NY 204, quoted in McNichol v. Flynn, 167 App. Div. NY 646.
[39]Nicolin v. Schneiderhan, 37 Minn. 63, 33 NW 33 S. Ct.
[40]Crandall v. Mary, 67 Ore. 18, 135 Pac. 188.
[41]U.S. — Finley v. Williams, 9 Cranch 164, 3 L. ed. 691; Ariz., Meyer-Clarke-Rowe Mines Co. v. Steinfield, 9 Ariz. 245, 80 P 400; Iowa, Currier v. Jones, 121 Iowa 160, 96 NW 766; N.Y. Drew v. Swift, 46 N.Y. 204.

and distances according to White v. Spahr, 207 Ga. 10, 59 SE 2d. 916 (1950). Although it has sometimes been claimed that monuments must be visible, there are any number of situations where they are not; i.e., iron pipe set for corners of subdivisions; monuments set for street center line intersections below the pavement; points set in farm lands sufficiently deep to be missed by the plow and harrow; utility pipelines and others. A recent court case established an underground pipeline as a monument. In particular, it said, "As a monument, its location takes precendence over the course-and-distance description set out in the 1941 instrument."[42]

In another instance, a stone was found and used as a sectional monument, fences were built on lines to and from it, and the fences had been accepted in those positions over a period of time; but then along came another surveyor with more curiosity than anyone previously. He dug around with a shovel and found another stone with marks on it which conformed with the government field notes. This, then, was the monument that had character and background by recognition of the markings. This is an actual case where not only the fences had been built to and from the unmarked stone and had been accepted for property lines for years, but the stone also had been used as a control point for highway development — then the curious surveyor found the true section stone with the correct markings on it some thirty feet from the other one. Here is an excellent example of the fact that the first stone had no background and no character by which it could be recognized, but was given some undue acceptance — until the right monument was found.

The old question of "doesn't occupation hold?" hinges on the statement long held by courts: "facts that a *correct* survey will show." Here would be shown the true line developed from the marked stone and the relationship of the fence line to it. The line of title acquired from the patent would be to the true line and the area occupied by the fence would be open to settlement between the adjoiners or possibly to a quiet title action.

Sometimes monuments are claimed to be indicative of a line described in a deed when in fact they do not conform with the calls, fit a background of history, or meet with acceptance by others than the claimant. In a case concerning the claim of a strip of land adjoining a

[42]Czarnecki v. Phillips Line Co., et al., Missouri Court of Appeals, St. Louis District No. 1, Case Nos. 35731 and 35887, recorded April 9, 1975. 524 SW 2d. 153, Mo. App.

cut-out of a large interior lot of an old subdivision, the plaintiffs desired to quiet title to this strip on the basis of two old monuments on the west line of said strip. In the processing of the case, a map for exhibit was prepared by a licensed surveyor and showed the lines described in the plaintiff's deed. Furthermore, this map showed not only the strip of land adjoining the plaintiff's land but proved that the description in the plaintiff's deed was sufficient unto itself, referring again to the definite and certain line. Neither of the two old monuments could be proved to have been set by the subdivider's engineer nor for the plaintiff's deed; therefore, the courts disregarded the monuments, even though testimony claimed they had been in their positions for thirty to forty years. This case was appealed and upheld in the Appellate Court.[43] It again brings into focus many facts substantiated by other cases which point to the one superior fact that strict application of the deed description surpasses and is paramount against the use of any monument which does not have background, character, recognition, and acceptance in conformity with that description.

As one of the pertinent points relative to whether an engineer or surveyor should accept "a monument that has been used for a long time" for control of his survey, consider the fact that in a quiet title action a plaintiff may recover only on the strength of his own title and not on the weakness of defendant's title. Therefore, "Where there is acquiescence in a wrong boundary and the true boundary may be ascertained by the deed, the acquiescence is treated both in law and equity as a mistake and neither party is estopped to claim to the true line."[44] This true line means that which is described in the deed.

The art of just finding monuments is not in itself sufficient. The need to understand the relationship between them as well as to know the rules of how to develop points and lines from them is a necessary adjunct to the finding. On a map submitted for the subdivision of that portion of the North half of the Southwest quarter of Section 15 T___ S, R___ W in the______ Meridian lying easterly of the Highway, four section monuments were shown as found at the east ¼ corner, the south ¼ corner, the southwest corner and the northwest corner of the section and the position of the west ¼ corner was established from two bearing trees. See Figure 53a. The east-west quarter section line was measured and divided in half to establish the center of section from

[43]Williams v. Barnett, 135 Cal. App. 2d. 607, 287 P2d 789.
[44]Janke v. McMahon, 21 Cal. App. 781.

which a line was drawn to the south quarter corner setting the east line of the southwest quarter and therefore the east line of the tract.

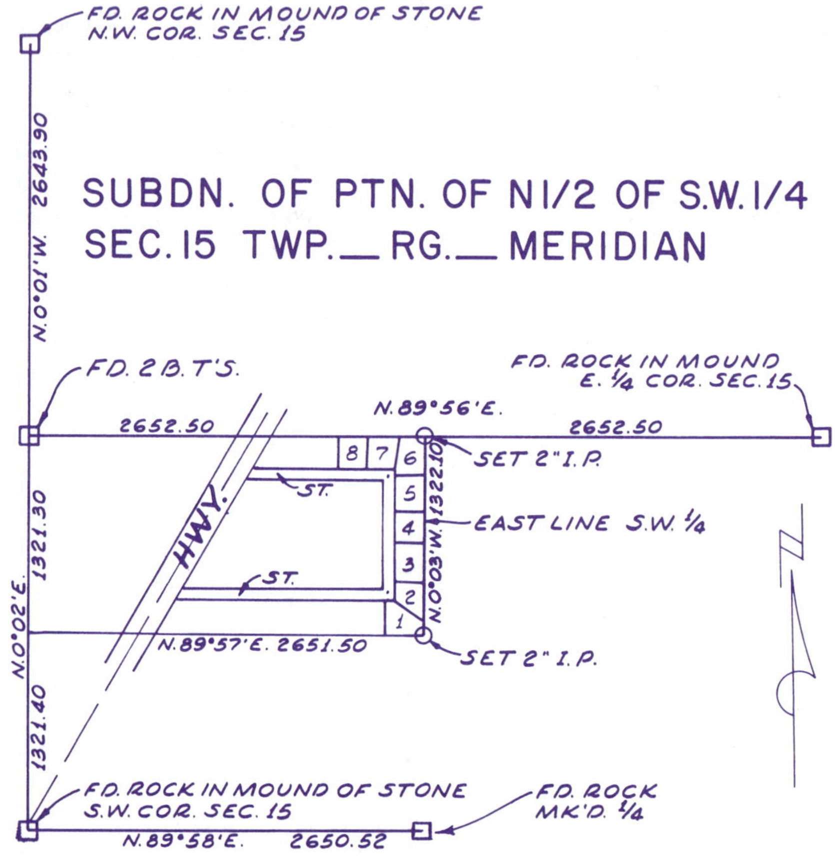

Fig. 53a

Following the recordation of this tract map, the owner of the southeast quarter of Section 15 decided to subdivide his land and requested his surveyor to proceed with the field work and mapping. The next drawing shows that his surveyor found, in addition to the monuments shown on the tract map, the northeast corner of the section from which he set the north quarter corner. When he followed the rule in the Manual for establishing the center of section by intersecting straight lines between opposite quarter corners, he discovered that his north-south quarter section line cut through the east row of lots in the recently recorded subdivision, which meant that portions of them encroached into his southeast quarter. See Figure 53b.

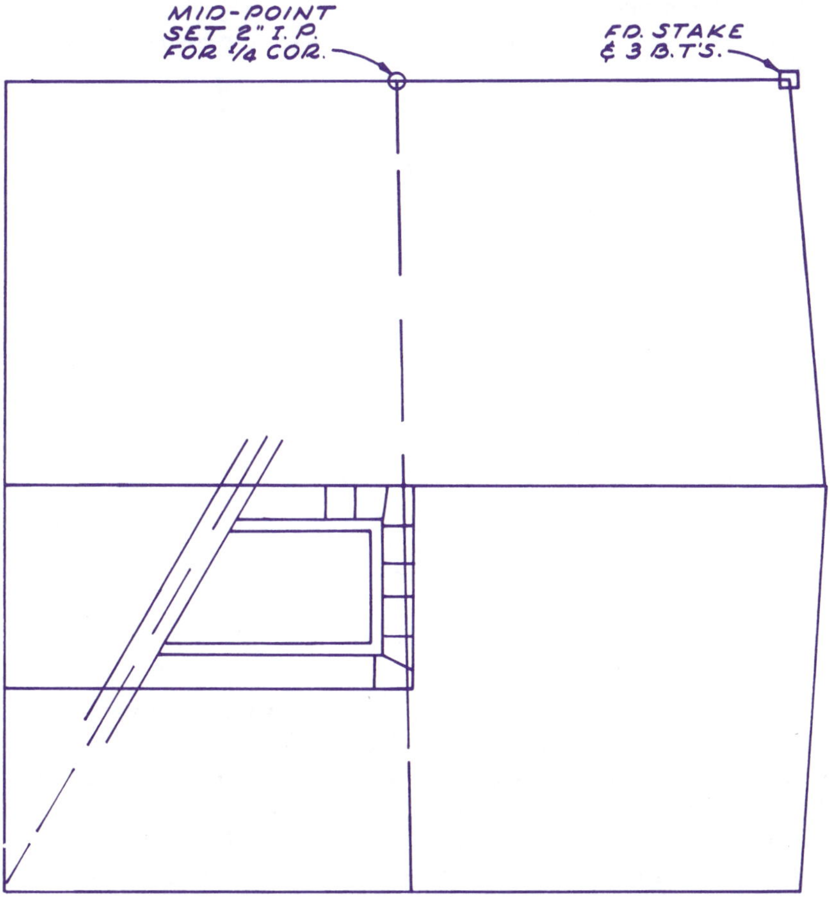

Fig. 53b

The conclusion is obvious. The first surveyor did not follow the rules. *He* should have developed the north-south center-of-section line instead of splitting the east-west center-of-section line distance.

Another case of encroachment arose because two surveyors accepted a fence post as the corner common to Sections 7, 12, 13 and 18. The tract map on the south filed by the first surveyor showed the line between Sections 12 and 13 as north boundary of the tract and the post as the section corner and northeast corner of the tract.

The second surveyor followed in the footsteps of the first one and filed his tract map on the north showing the same section line as his tract's south boundary, and the post for the section corner. See Figure 53c.

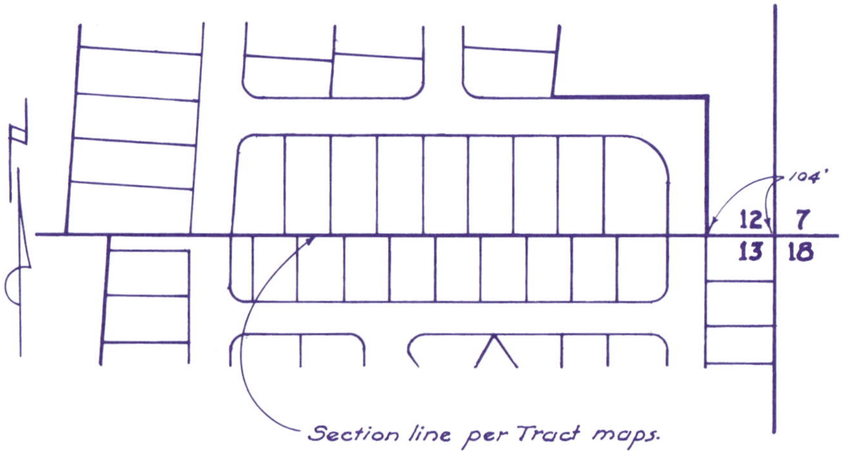

Fig. 53c

A third surveyor went hunting and found two spruce bearing trees according to the government surveyor's field notes. From these he established the correct position for the corner common to Sections 7, 12, 13 and 18 at 49.4 feet northerly of the fence post used by the other surveyors as shown in Figure 53d.

In another situation, the architect made a request for the surveyor to set the four corners of his client's property and furnish a map showing the dimensions. Having acquired said map, the architect proceeded with his plans, making maximum use of the available area and designed the buildings to be set only two inches from the property lines. After the contractor completed the structures, the adjoining owner decided to also build and ordered a survey of his property. The

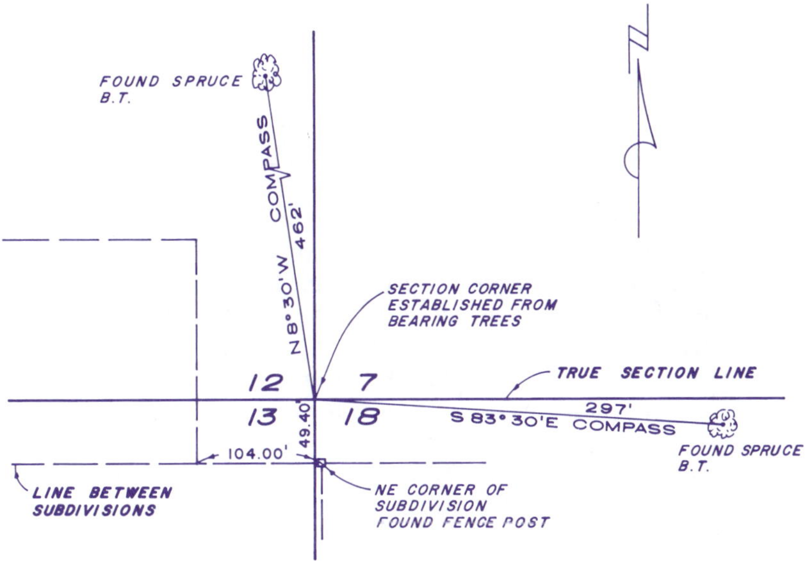

Fig. 53d

map submitted showed that the recently erected building encroached onto the rear of his land almost four inches. Inquiry to the contractor and surveyor of the first property revealed the following facts:

1. The surveyor found three monuments at the rear (NE) corner and, in addition, set his own iron pipe.
2. The surveyor showed only his monument at said rear corner.
3. The contractor saw only one iron pipe at said rear corner and he ran his string line from it to the iron pipe at the front (SE) corner. He did not dig to see if there were any other monuments.
4. The iron pipe used by the contractor was about six inches east of the one set by the surveyor, thereby putting the building four inches over the line.

Since a contractor is not supposed to be a surveyor (and dig for monuments), it is up to the surveyor to make his points definite, both on the ground, on the map and in the description.

Bear this in mind also that finding a monument that conforms with the description of the land next to your survey but leaves a gap between the two parcels gives you no privilege nor authority to annex that gap to the land of your client. Rather, it imposes an obligation upon you to inform your client of the facts so that he may consider whether he desires to make proper arrangements to acquire title to the gap or to stay out of it.

The statutory rule for preference of monuments over measurements is for use as a crutch, and for construing a description, and does not mean that the court may go outside the instrument to find a monument not mentioned therein as a means of making the description in the instrument fit the monument.[45] For instance, in the citation, "beginning at a point on the east line of Section 35 north 1321.13 feet from the southeast corner thereof, being the northeast corner of the southeast ¼ of the southeast ¼ of said section as shown on a Record of Survey Map filed in the County Recorder's Office," and where the total distance is also shown for the whole east line of the southeast ¼ as 2644.72 feet, which distance should be used, 1321.13 feet or the split which equals 1322,36 feet? Because the citation gives both the 1321.13 feet and the ¼-¼ corner reference which by split would be 1.23 feet further north, one needs the corroboration, the crutch as it were, of some outside thing to help make a decision. For this purpose, an investigation and a check measurement revealed the fact that an iron pipe on said map 20 feet west of the 1321.13 foot point was only 0.06 feet north of its position on the map, which was a good indication of the particular distance intended by the writer of the description. Here is an excellent example of how a monument can give locative support for concluding the interpretation of a call but would not be in itself the finite control (that is, changing the 1321.13 feet to 1321.19 feet) because the iron pipe had not been mentioned in the description.

The specific reference to a map leads to another aspect of legal monument qualifications. "When there is a description by courses and distances, and another by reference to known monuments, the latter prevails; and a map of plan referred to (certainly if a public record) stands upon the same footing as a monument."[46] This is true, of course, only if the information on the map is correct.

In conclusion, when the scrivener or surveyor is conciously alert to the applicable value of monuments, he will use those which enhance the survey and the security of the description and ignore those which would be detrimental or which would introduce problems.

[45]Williams v. Hebbard, 33 CA 2d. 686, 689, 92 P 2d.
[46]Coles v. Yorks, 36 Minn. 388, 31 NW 353 S. Ct.

PROBLEMS

1. Give two definitions for "monument."
2. What are the two general classifications to which "monuments" are assigned?
3. Is a reference to an adjoining deed sufficient in itself to control the boundary?
4. Does a tie to an adjoining road as a monument carry fee title to the center of the road?
5. What kind of title does the owner of a lot in a tract shown on a map generally have in the adjoining street?
6. In a description of a part of a lot shown on a proper map, is it correct to assume measurements are to or from the center line of the adjoining street? Explain by court references and details.
7. Can a street without any lots adjoining it be a monument?
8. Is a canal or railroad a monument?
9. Can original monuments hold the position of a lot against the control of the street around the block? Explain.
10. Is every monument found by a surveyor acceptable for control? Explain.
11. If a monument is found near a line but *not cited in the description*, does it have authority to move the line over to it?
12. Does the timing of setting monuments before or some time after the writing of the description have any effect on the inherent value of those monuments?
13. Explain the difference between "replacement" and "relocation."
14. Must a monument be visible?
15. What are the three tests to apply to a monument before accepting it?
16. Is a "map" a monument? Explain how.

— NOTES —

Chapter 9

Streets

Ownership in land in some cases carries title to the center line of the street adjoining and in other cases it does not. This will be a study of various situations to learn the use and combination of words which in one form conveys or maintains the adjoining owner's fee title in and to the area covered by the street, while in another form, it cuts off such title. Where land has been subdivided, the lot or parcel normally carries the *underlying* title in the adjoining street to center line, unless it is a marginal street or has some other qualifications.

The creation of a strip of land for the non-exclusive use by people for road purposes may be by deeds, by dedication on a map, or by prescriptive use. In former times the county board of supervisors, or some such similar body politic, would, on petition from a group of landowners in a district, declare a road into existence over certain lands and then endeavor to acquire the deeds from the respective owners. Even if some were unobtainable, the road was built anyway and the county relied on prescriptive rights to establish their interest in it. Although by law in many areas, counties were permitted to acquire only an easement for public thoroughfares, in later years the laws have been changed to allow any political unit, city, county or state, to acquire fee title for land in public use. Such title may be obtained by the common deed form or, in rare cases, it may be transferred by use of the proper form and words of grant in the owner's certificate on a map duly recorded.

If the strip is to be limited in use, it may be acquired as an easement or a fee through a document setting forth the limitations; otherwise, the condition or limiting qualification may be designated, such as, "for road purposes." When a street is created by an easement over a strip along the boundary of subject property, the fee title in and under the street remains in the owner or his assignees and to the original boundary.

When land is described by metes and bounds subsequent to the creation of the street, the format of the words will determine the extent of title in and to the street. When shown on a map, recorded or not, it is not to be generally assumed that the street is included within the lot. Even if a map carries some such note as, "Distances and acreages are measured to center line of adjoining streets," it does not move the lot line and the street is *not* by these words included with the lot. The case of Earl v. Dutour, (1919) 181 C 58, 183 P 438, (which in turn referred to an Indiana supreme court case) clearly sets forth the fact that a lot and a street are two separate and distinct terms; this summarized principle refers to usable land. The other qualifications have been reviewed in Chapter 8.

Figure 54 shows a Lot 12 cut into five parcels fronting on HILL ST. being a dedicated street within the tract.

The following descriptions of the parcels sold out of Lot 12 illustrate the effect of words as to the conveyance of grantor's title to portion of street adjoining.*

Parcel 1: The North 50 feet of Lot 12. This is sufficient to include grantor's title to portion of west half of the street adjoining said lot.

*From Proceedings of the California Land Title Association Conventions May, 1942 and June, 1952.

Parcel 2: The description in the conveyance goes "to the Westerly line of Hill Street and thence Southerly along the Westerly line of said Street." It has been held in California that this language is an expression of an intent to convey title only to the side line of the street so that the grantor's title in the street is not transferred. (Warden v. South Pasadena Realty & Improvement Company, 178 Cal. 440; Severy v. The Central Pacific Railroad Company, 51 Cal. 194.) The same result follows where the description goes "to the Westerly line of Hill Street and thence Southerly along said line." (Berton v. All Persons, 176 Cal. 610, 614.)

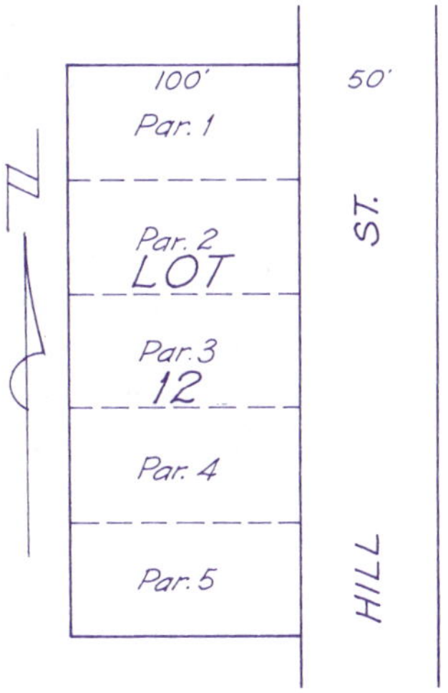

Fig. 54

These decisions in California follow the majority rule in holding that a call expressly for the side line of a street excludes the street from the grant. (11C.J.S. 584). The rule is one of strict interpretation established first by early decisions in New York and it is hardly consonant with the declared policy that the presumption — that is, that the owners of lands adjacent to a road own to the center thereof — is "highly favored by law." In contrast is the liberal or "New Jersey rule" that nothing short of express words of exclusion will exclude the street, such as "except the portion in the street." (Salter v. Jonas, 39 N.J. Law 469, 23 Amer. Rep. 229). The theory upon which the liberal rule is based is sound; the principals in a transaction rarely have any understanding as to whether the street is to pass or not; the general notion is that the street should belong as an appurtenance to the contiguous lot; and normally a close and literal meaning given to particular words in a description does not give affect to the actual intent of the grantor but serves only to defeat the policy of the law. There is much to commend the New Jersey rule, and conceivably, our courts today would follow the rule were it not for what seems to be a rule of property to the contrary.

Parcel 3: The description in a deed goes "100 feet to Hill Street, thence at right angles *along the street*." The distance given would extend only to the side line of the street. It has been held that the street is thus made a boundary and that the title passes to the center of the street, regardless of the fact that the distance given is only to the side line. (In Moody v. Palmer, 50 Cal. 31, the Court said: "It is well settled that land described in a deed as bounded by a public highway or street will be considered as extending to the center of the street or highway, unless it clearly appears that it was intended to make a side line instead of the center line the boundary. The highway is a monument and in legal contemplation, the *thread of the highway* is the monument , unless a contrary intention clearly appears.")

Parcel 4: The description begins at a point or monument in the West line of Hill Street and "*thence along said street,*" or the call is to "the West line of Hill Street" and "*thence along said street.*"

It will be noted that the call is "along the street," not "along the West line" of said street nor along "said line."

The authorities are not in accord on the question of whether descriptions which tie to the side line and then go along the street rather than along the side line include grantor's title to the center of the street. Some authorities hold to a technical rule which excludes the street (King's County Fire Ins. Co. v. Stevens, 87 N.Y. 287 and other cases cited in the annotation at 2 A.L.R. 21) while the more liberal rule followed by other courts holds that the fee passes to the center of the street under descriptions of this kind. (Hensley v. Lewis, 278 Ky. 510, 128 S.W. 2d. 917, 123 A.L.R. 542, annotation, stating that the weight of authority supports the rule that a description which begins "at the edge of county road; thence running with the road" carries the title to center of road.) It is probable that the courts in California would follow the liberal rule in this instance. This is indicated by the previously mentioned case of Moody v. Palmer, which also held that a description by metes and bounds which calls for a line beginning at the "*corner*" of two streets, thence running "*along*" one of them does not prevent the grantee from taking to the center.

Parcel 5: The description goes to the "East line of Lot 12" and "thence along said line."

Under the rule in Anderson v. Citizens Savings and Trust Co., 185 C 386, 393, that the *street shown on the map is a part of the lot* AND THAT the *technical* lot line is the center of the street, it seems proper to hold that the description carries to the center of the street, unless the grantor clearly expresses, in the granting instrument, a contrary intent.

Other descriptions deemed sufficient to carry the street are: "bounded on the North" by a street (Watkins v. Lynch, 71 Cal. 21) or "bounded on the North by the *south side*" of a street (Brown v. Bachelder, 214 Cal. 753).

In general, these rules should apply to all transfers of land but a federal district court held that in an eminent domain and declaration of taking of land described by metes and bounds, such presumptions may not apply (U.S. v. 5.324 Acres of Land, 79 F Sup. 748). On the other hand, a condemned lot shown on a map with a boundary has been held to include the street (Cleaver v. Board of Education, 248 NW 629).

The essence of this is twofold: in reading such descriptions concerned with streets, analyze them to know whether title has been carried into, or cut off from, the area used in common with others. In writing descriptions, be certain that title in the street has not been taken away, nor added to, that which the owner already possesses. This may require help from an attorney or a title company.

Dedications

Any dedication of land, whether it be for parks, streets, alleys, walks, bridges, wharves, etc., expresses an intent for the land to be accepted and used for that public purpose. It may be accomplished by either statutory or common law. The dedication of road ways on maps is a form of easement for the benefit of all ownerships covered by the map, and the public too if so specified, with the underlying interest or fee title in the street shared by adjoining owners in whatever proportion would pass by a conveyance of their property. The respective participation or cut-off would depend upon prior boundary title.

Vacation or Abandonment

The question of the exact outline of those areas in the street which would be apportioned and attached to their respective parcels is seldom raised until the street is vacated. There appear to be no settled rules of law detailing the outline of the areas concerned in this matter.[47] However, one court case, State v. Superior Court (Wash. 1918) 173 P 186, 189, held that the division of the vacated street should be based on the prolongation of the side lines of the lots as shown on the subdivision map. The adaptation of this method was

[47]Ogden's Revised California Real Property Law, Vol. 1, 640.

also decided upon in Oklahoma.[48] In some states, such as Kansas, the reversion of areas in a vacated street is based on proportional frontage. In one California case,[49] it was held that the conveyance of a lot by reference to a tract map, showing the bounding street, carried with it the grantor's fee to the center of the street even though the street had been vacated. A problem arises if the lot lines are acute and cross each other before reaching the center line; there would appear to be not only an inequity of distribution but a question as to whom would acquire the crossover area 'x'. See Figure 55.

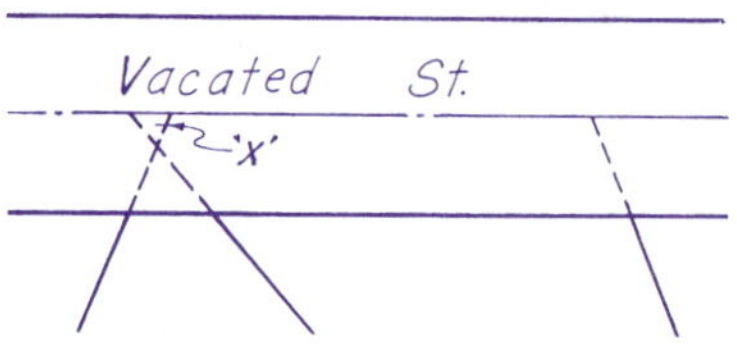

Fig. 55

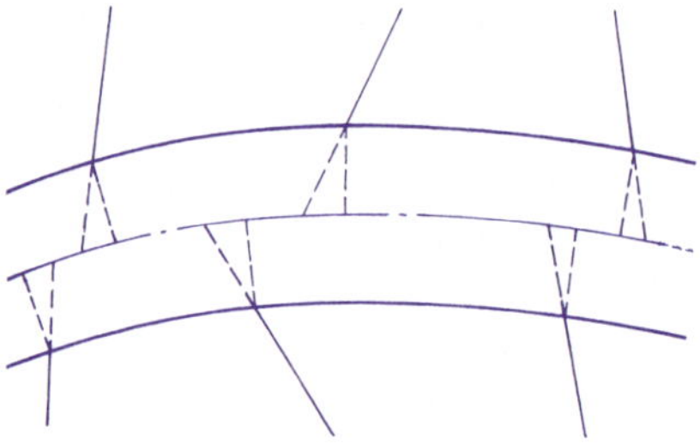

Fig. 56

The matter of apportionment prevails with many different configurations. In the case of lots on a curved street, will the division be based on the prolongation of the side lines or on the radial lines of the curve to the lot corners? See Figure 56. It depends upon what the applicable code says, if anything, or it will be up to the courts.

Problems also arise when the lots in a tract abut a right angle street or share a deadend street. See Figure 57.

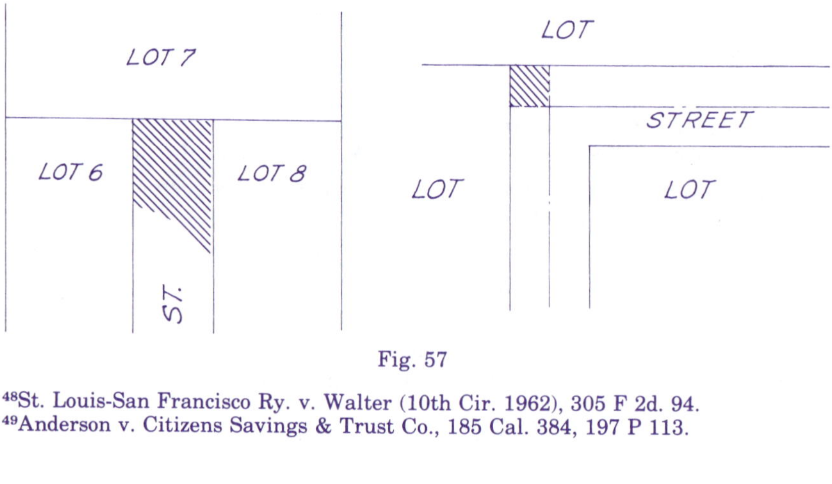

Fig. 57

[48]St. Louis-San Francisco Ry. v. Walter (10th Cir. 1962), 305 F 2d. 94.
[49]Anderson v. Citizens Savings & Trust Co., 185 Cal. 384, 197 P 113.

Another version of the right angle theory would not, however, satisfy the requirement of frontage for Lot 15 on the new street alignment, as shown in Figure 58, if the solid triangle reverted to Lot 23 in accordance with that right angle theory. Logic and practicality would divide it between Lots 14, 15 and 16 for equitable access.

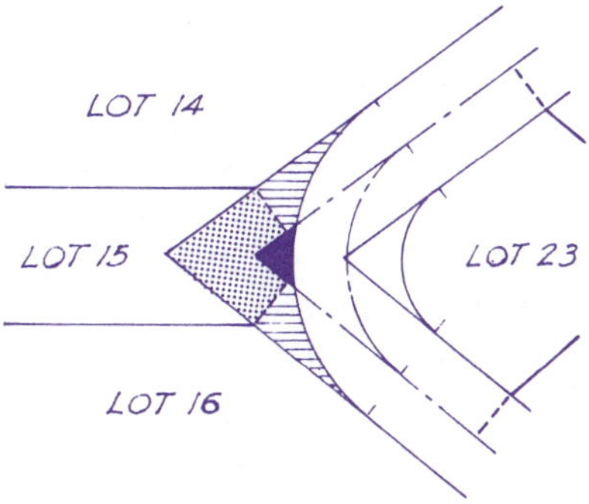

Fig. 58

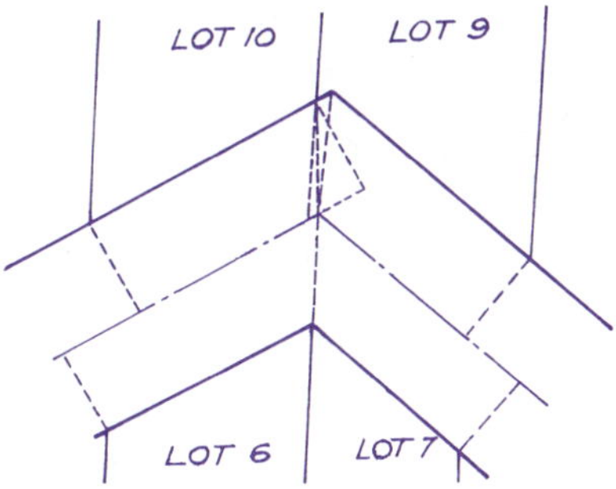

Fig. 59

In the situation of an angle street, as shown in Figure 59, where the normal process uses the bisection of the angle for division of interest in the street, where should the dividing line between Lots 10 and 9 be placed? On the bisector, at right angles, on the prolongation or on a compromise line from the lot corner to the bisection point on the center line?

A street made in whole from and along the margin of grantor's land is called a marginal street. When street is margin to a row of lots and is itself boundary to the tract, *all* of its underlying title may vest in the adjoining lots within that tract, but it is necessary to trace the title back to the conditions that existed at the time the marginal street was created to determine which ownership created it and if any adjoiner had any rights in and to it. See Figure 60.

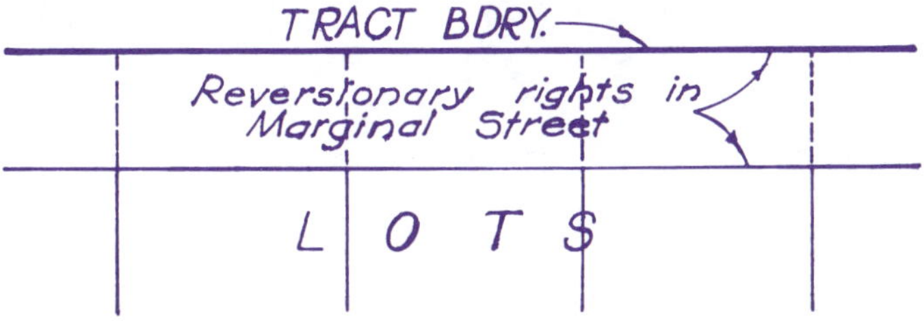

Fig. 60

Any boundary of a subdivision, in whatever way it lies in relationship to the streets or lots, limits the title interest of those lots to the area within that tract. Interests outside the subdivision can have no reversionary rights in and to the street within the tract, or vice versa. See Figure 61.

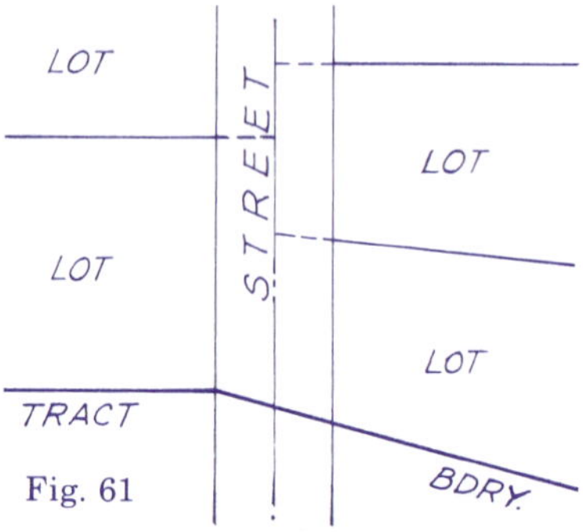

Fig. 61

With these many possible combinations and no finite rules established beyond the very few court cases, the scrivener who is confronted with having to write a description of a vacated part of a street might do well to follow the form used by some title companies; viz:

> Lot ___ and that portion of 'A' Street vacated (date) by Resolution No. ___ that would pass by a conveyance of said lot, of Tract No. ___ in the County of ________ etc.

This leaves the definitive decision up to a court based on the circumstances and the law, including also the segregation of fee and easement if there is any such question.

If a street or right-of-way is acquired by individual parcels put together to form the strip, then any reclamation of such parcels due to abandonment would follow the same lines as granted. See Figure 62. The hook lines indicate the associated parcels granted which would return to their respective owners at the time of abandonment.

It must be remembered that any vacation or abandonment of publicly-held land (whether acquired by an easement deed, a fee deed

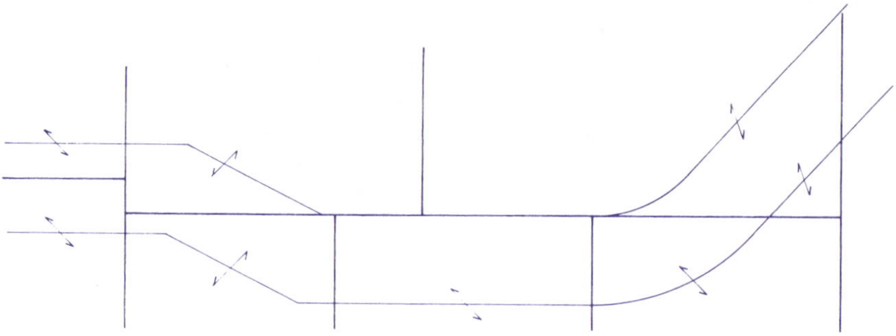

Fig. 62

or dedication on a map) only abandons the *public* rights; private rights remain intact. The foregoing examples have pointed up some of the problems in the distribution of the areas of those rights.

Reversion of Title

Although some think of vacation and abandonment as reversion of title to the owner, it is really not so because the status of the title remains the same before and after vacation with only a change in the use of certain rights. Reversion of title is used in the sense of reclaiming rights to land after the condition has occurred that was imposed upon the property ownership at the time of conveyance or after abandonment by the grantee, such as when a railroad company ceases to use its right of way for railroad purposes, as specified in the deed. This use of "reversion" is technically inaccurate; instead, the grantor's right to terminate that estate on happening of the condition is sometimes called a possibility of reverter; in Parry v. Berkeley Hall School Foundation,[50] it was defined as a power of termination and right of reentry and according to Johnston v. Los Angeles,[51] it is transferable in the same way as estates in land.

An interesting situation developed when one owner gave a deed to a railroad company for a strip of land across his property and included a reversionary clause in case of non-use for the specified purpose. Subsequently, he sold all the land on one side of the railroad strip to a developer who subdivided it with a marginal alley against the railroad strip. In later years, the railroad quit its use of that strip but the grantor had to go to court to regain his physical possession of it. In his prayer, he claimed his right to the entire width of the strip but the court decided that in the sale of the adjoining land, the developer (grantee) acquired the reversionary rights in and to the half of the railroad strip bounding the subdivided land. Result: Because after-acquired title passes on to the current owner, each lot along the marginal alley acquired a portion of the abandoned railroad right of way opposite the lot. See Figure 63.

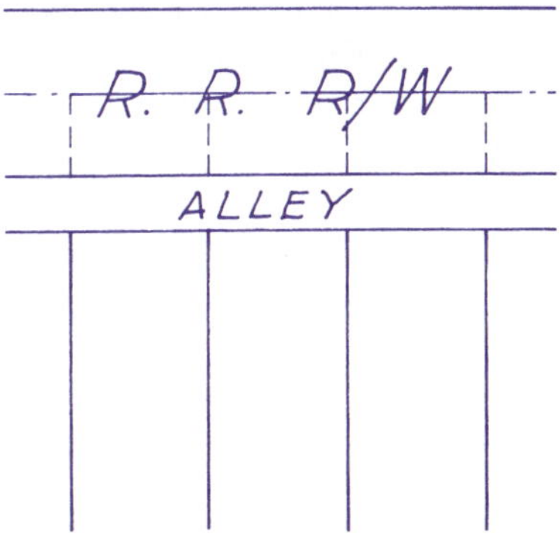

Fig. 63

[50]10 C 2d. 422, 74 P 2d. 738.
[51]176 C 479, 485, 168 P 1047, citing Civ. C. 1046.

There are many variations in the ultimate outcome of the right of reentry or power of termination dependent upon the particular wording in the document as evidenced by the court cases involved in this subject matter; they make interesting reading.

A different aspect of reversion has to do with special interests in streets in a subdivision when the subdivider wishes to revert the tract back into acreage for some other use. If for example any utilities have been placed in those streets for the benefit of adjoining lands, the subdivider may not disregard the beneficial rights obtained by them. The vacation of any such streets must except the easements of existing utilities or private rights which may have been established since the official creation of the tract.

Description Control

There are several types of control in the description of streets, alleys, ways, etc., with which you will be confronted both in reading and writing them. The most common, of course, are the reference to a showing on the map and the center line control of a strip of land. In contrast you may have a survey line control which is not a center line because there will be one width of footage on one side and a different width on the other side, or, the entire strip may lie on only one side of the line. In the case of irregular widths, perhaps they will be related to, and at various distances from, an offset line. More detail on this will be given in Chapter 12.

Questionable Maps

This has to do primarily with older maps made before regulations and formats were prescribed by law because they did not make specific delineations on the maps as to the use of the streets.

Today, in many areas, it is required that in the owner's certificate on the map, the status of the streets, alley, ways, etc. be declared as "dedicated," or not, or "private" or as having limited access, etc.

It has been said that the making and recording of a map showing streets constitutes an offer to dedicate those spaces designated as streets to the public as streets.[52] Dedication consists of an offer and the acceptance thereof. "A dedication without acceptance is, in law, merely an offer to dedicate and such offer does not impose any burdens nor confer any rights, unless there is an acceptance."[53] The acceptance may be formal or implied but it must be unequivocal.

[52]Tischauser v. City of Newport Beach, 225 Cal. App. 2d. 138, McKinney v. Ruderman, 203 Cal. App. 2d. 109.

[53]County of Inyo v. Given, 183 Cal. 415.

When the older maps say nothing about the streets, there may be certain prescriptive rights established over time to uphold the status of their use as streets. The question of a grantee, to a lot referring to the map, having title in the street adjoining is dependent upon two conditions: (1) The wording in the deed of conveyance (as discussed supra), and (2) The time relation between the creation of the street, both on the map and physically, and the sale of the lot.[54] Although in cases cited in New York and Illinois, it was held that even if the buyer obtained a lot shown on a map as abutting a dedicated street, he did not acquire title in fee to the center of the street unless that dedication had been accepted by the public. Contrariwise, Johnson v. Arnold[55] and Amerman v. Missouri, etc. R. Co.[56] stated that such ownership "was not dependent on whether the public accepted the dedication or abandoned its rights thereto."

Another aspect concerning the acceptance of the dedication of streets offered on a tract map is the fact that the political entity involved may not want to have the responsibility of maintaining the street and on that premise alone will fail to accept the dedication in order to avoid such responsibility.

In any event, the problem involved here concerning description writing is the retention of the owner's rights, if any, in the street.

PROBLEMS

1. Will "distances and acreages measured to center of adjoining streets" cited on the map change the lot lines from side line to center line of streets?
2. Give the principal form of wording that cuts off title to the street area other than by outright EXCEPTION.
3. Give one of the forms of wording of a description to the side line of a street which will carry title into the street area.
4. Is a dedication on a map of a public area a fee or easement type of status?
5. Does the vacation or abandonment of a street remove the private rights of others than the respective owners fronting on the street?

[54]Trowbridge v. Ehrich, 116 App. Div. 457, 101 N.Y.S. 995 (mod. 191 N.Y. 361, 84 NE 297).
[55]91 Ga 659, 18 SE 370.
[56](Tex. Civ. A.) 182 SW 55.

6. Does a vacation of a street or alley give the adjoining owners the right to build onto it?
7. Describe a marginal street.
8. Describe "reversion of title."
9. Can the vacation of a street or alley, etc. ignore the rights of easements created therein?
10. Name three kinds of description control for strips of land.
11. Write two descriptions of the easterly portion of Parcel 1 shown on the map on page 9.13.
    a. As a part of the Parcel (with reference to the map) lying east of the ℄ of the 10-foot sewer easement.
    b. As a metes and bounds description written to the *side line* of Whittier Boulevard in such a way as to include title in the street and lying east of the ℄ of the 10-foot sewer easement.

    In both cases, show subject to specific encumbrances.

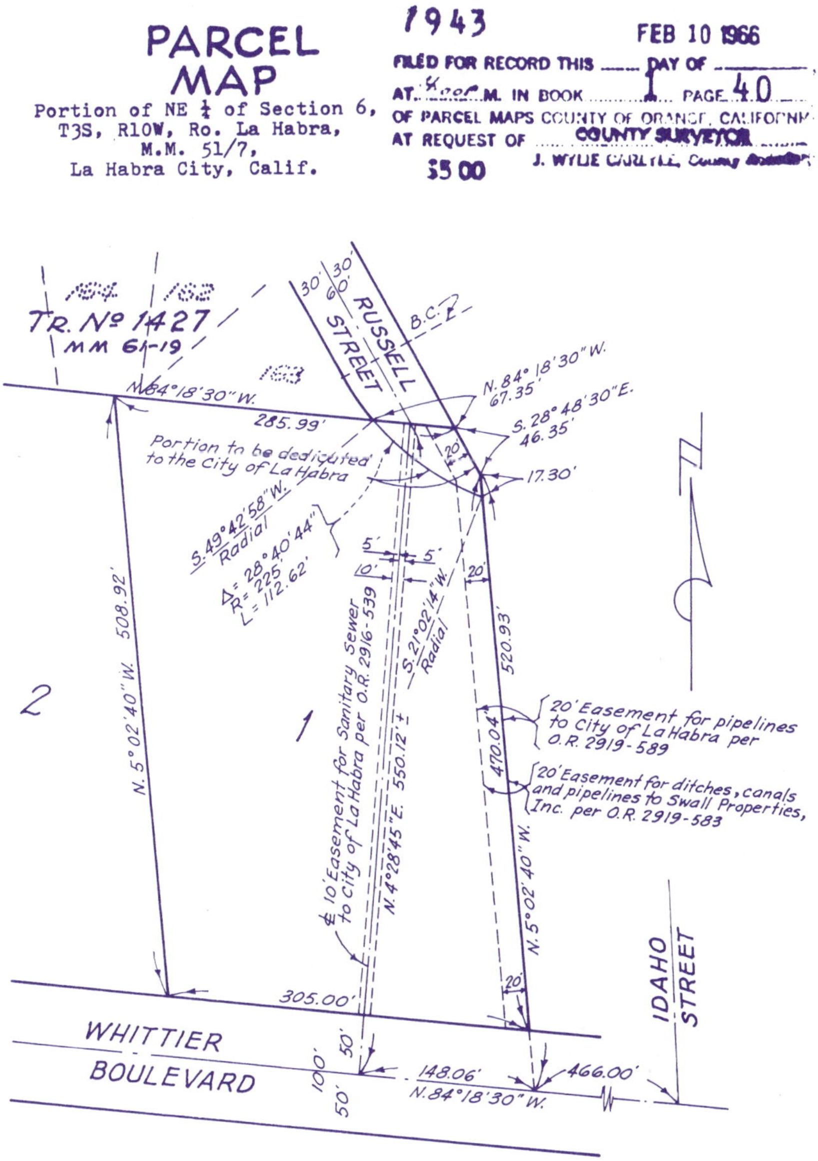
PARCEL MAP
Portion of NE ¼ of Section 6, T3S, R10W, Ro. La Habra, M.M. 51/7, La Habra City, Calif.
1943
FEB 10 1966
FILED FOR RECORD THIS ____ DAY OF ________,
AT ____ M. IN BOOK 1 PAGE 40
OF PARCEL MAPS COUNTY OF ORANGE, CALIFORNIA
AT REQUEST OF COUNTY SURVEYOR
J. WYLIE CARLYLE, County Recorder
$5.00
TR. No 1427
MM 61-19
RUSSELL STREET
B.C.
N.84°18'30"W.
285.99'
N.84°18'30"W. 67.35'
S.28°48'30"E. 46.35'
17.30'
Portion to be dedicated to the City of La Habra
S.49°42'58"W. Radial
Δ=28°40'44" R=225' L=112.62'
508.92'
N.5°02'40"W.
2
1
℄ 10' Easement for Sanitary Sewer to City of La Habra per O.R. 2916-539
N.4°28'45"E. 550.12'±
S.21°02'14"W. Radial
520.93'
470.04'
20' Easement for pipelines to City of La Habra per O.R. 2919-589
20' Easement for ditches, canals and pipelines to Swall Properties, Inc. per O.R. 2919-583
N.5°02'40"W.
305.00'
WHITTIER BOULEVARD
148.06'
N.84°18'30"W.
466.00'
IDAHO STREET

— NOTES —

Chapter 10

Occupation vs. Title

The status of occupation exists in several ways but predominantly in full belief and satisfaction of good title. For example, when you buy a house and lot described on the deed as Lot 8 in Block C of the Newhaven Tract, etc., etc., or you buy a farm designated on the deed as the SE 1/4 of the SW 1/4 of Section 10, T 3 S, R 7 E, etc., etc., in such cases you consider that the boundaries are OK, there are no gaps or overlaps and there are no encroachments of walls, fences or buildings into or out of your property so acquired.

Acquiescence

The obverse conditions of occupation manifest as misalignments of physical things in their relationship to lines of title and there are several conditions of status of which perhaps the most common is acquiescence. This often results from adjoining owners having bought the property after the problem was created and either they are unaware of the situation or if they have been apprised that there may be a problem, they are accepting and holding it in status quo rather than to make a fuss over it.

The next step in this picture is taken when either party determines by one way or another the true facts in detail at which point of knowledge, acquiescence as such ceases to be and the condition turns into a problem to be solved. This change of attitude does not stop either party from claiming to the true line — that is, the one described in the deed — and if it is a definite boundary, it must govern.[57] Again, the fact is cited that "The boundary is considered definite and certain when by survey it can be made certain from the deed."

To always have such definitiveness that there could be just one and only one position for each boundary line would be Utopia, but from various and sundry causes, some lines become bound in a web of confusion. Therefore, in order to devolve upon some general groove of thought, we will consider that a line which determines the demarcation of property is true and correct and carries the weight of decision when it has the preponderance of weight by delineation in words and numbers which fits with all correlated parts of the whole of each description on each side of that line. Multiplicity is sometimes complicity — and it certainly is when more than one line is claimed for ownership. Mere claim, however, is not sufficient unto itself.

Multiple Lines

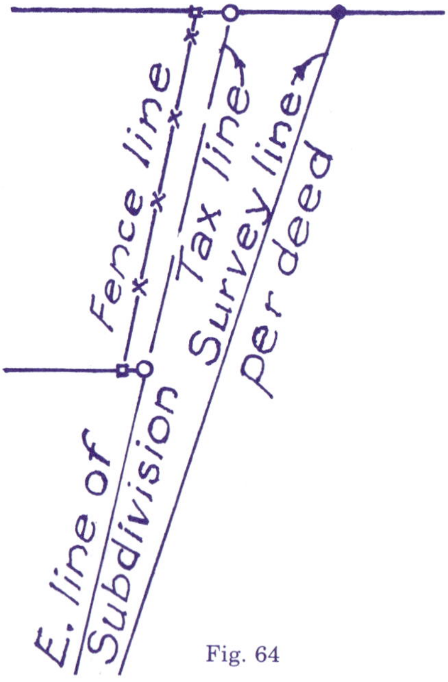

Fig. 64

Let us consider the example shown in Figure 64. Three different lines have been created and claimed as boundary between two holdings of title; a fence line, a tax line, and a survey line. The fence line has been in existence over twelve years and has been acquiesced in by various successive owners who, for many years, were unaware of the possibility that it was not coincident with their true line. The tax assessor's office has been collecting taxes all this time on the basis of a line produced northerly from the east line of an old subdividion on the South, and which line was nearly parallel with the fence but a few feet east of it. The more recently surveyed line is some 23 feet east of the fence and was brought into focus by reason of the fact that the land was becoming more valuable and a proposed transaction on the east side was to be based upon the survey for determining the payment of so many dollars per acre.

In the case at hand, the moment the adjoining owners on the west realized that the line to which they had been paying taxes was a few feet east of the fence, the value of the fence as a line dropped in favor of the tax line. Then when these owners soon after learned that their line of title as developed by the survey was some 23 feet east of the fence, the value of that tax line to them dropped in favor of the title line. And conversely, the owners on the east intensified their evaluation of the tax line over the title line because they had been paying taxes thereto for many years. Now it appears that the fence has been eliminated from the problem and well it could be in consideration of the fact that it was only for stopping cattle because "where acquiescence in a fence is as a barrier and not as a boundary line, no agreed boundary is established."[58] On the other hand, the payment of taxes on land NOT included within the description is NOT effectual to complete title by adverse possession.[59] And, "the payment of taxes by the adverse holder on a designated parcel of land is not effectual to complete a prescriptive right to land not included within that designation."[60] Also, in California, such title is not a "marketable title," i.e., one "free from reasonable doubt and fairly deducible from the record," which a purchaser can be compelled to accept (Gwin v. Calegaris, 139 Cal. 382) until the title is established by judicial proceedings against the record owner, even though "no title can be better or more absolute" than that acquired by adverse possession according to Woodward v. Faris, 109 Cal. 12.[61] These decisions would seem to knock out the tax line as a claimed boundary until or unless a quiet title action is instituted and a judgment rendered in favor of the occupants on the east.

In certain areas, a claimant to adverse possession must have paid taxes prior to and continuously during the processing of his claim,

[57]Hartung v. Witte, 59 Wis. 285, 18 NW 175.
[58]Chandler v. Hibberd, 165 Cal. App. 2d. 39, Talmadge v. Moore, 98 Cal. App. 2d 481.
[59]2 Cal. Jur. 2d. Sec. 72-173.
[60]Eberhardt v. Coyne, 114 Ca. 283, 46 P 84.
[61]Ogden's Calif. Real Property Law, 5.28.

while in other areas, a color of title and payment of taxes for seven years is admissible for claim of title by adverse possession but the color of title must be first. Money paid for taxes on disputed areas can constitute a separate claim by itself.

It appears that by these circumstances and without the benefit of a quiet title action, the limit of the land to which either party has the right to convey (outside of to each other if they are to establish an agreed line) is not the fence or the tax line but only the line of title as described in the description in their deeds, because "where there is acquiescence in a wrong boundary and the true boundary may be ascertained by the deed, the acquiescence is treated both in law and equity as a mistake and neither party is estopped to claim to the true line."[62]

The surveyor on the job should show all lines on his map properly identified.

When knowledge of the facts replaces blind acquiescence, there are usually two courses of procedure when use of the same line is continued. If the adjoiners are satisfied with the acquiesced-to line, they can establish it in a legal manner by an agreement. If one party to the line disagrees but the other desires to hold it, then it may be necessary for the latter to claim under adverse possession through the proper applications of law. If an agreement is consummated and legally established, any former condition or status of adverse possession due to the difference between the "agreed" line and the "true" line is obliterated.

The acceptance of lines supported by mutual agreement through private settlement by coterminous owners is blessed by the courts in that it precludes useless and time-consuming litigation.

Agreement Line

As to whether that agreement concerning a mutual boundary can or should be oral or written hinges on two other conditions. One is a statute based on old English law about fraud which requires, among other things, that agreements affecting interests in land in order to be valid must be in writing. The other concerns third party rights where a buyer or lessee for example would have no cognizance of such an agreement. In either case, the third party must be informed either verbally, by presentation of a copy of the written agreement, or by

[62]Williams v. Barnett, 135 Cal. App. 2d. 607, 612, Pra v. Bradshaw, 121 Cal. App. 2d. 267, 263 P. 2d. 52.

accessibility to it through public records. Furthermore, an agreement may be only that and in fact make no conveyance of title either way across the disputed area that is between the agreed line and the one that has been on record, such as are shown in Figure 65. If there are not proper words of conveyance in the agreement, no title passes between L and M, and therefore, any subsequent sale of the land would not carry title to or from the "agreed" line but would remain only as shown by the record line. The alternative to having words of conveyance in the agreement is the execution of separate documents between L and M, provided they hold fee title, which would convey their respective areas to the "agreed" line, thereby establishing it legally. This operation is sometime called the "practical location" of a boundary.

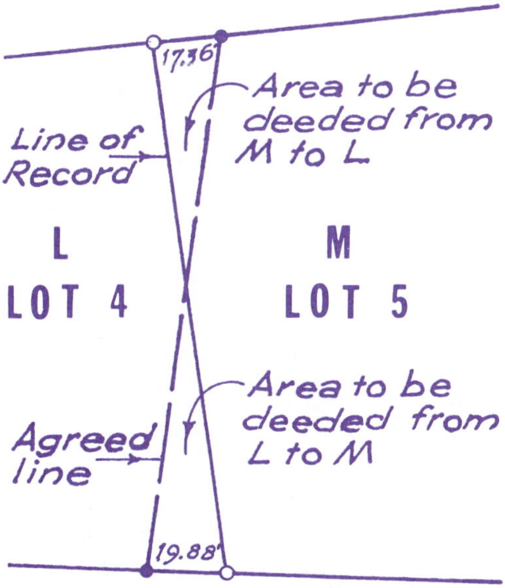

Fig. 65

Another situation evolved from an actual agreement line established by a document setting forth a mutual understanding between two certain adjoining parties and with specific words of conveyance to each other on each side of the agreed line. The first impression seems to indicate a perfect deal — but — consider the facts indicated in

Figure 66. A, B and C held their adjoining titles with properly described common boundaires. The easterly fence line, though crossing all three areas, had been used as the line of demarcation for many years between B and C who entered into said agreement and with proper words of conveyance established the fence line as their boundary. Note, however, that C's occupation to the fence included a triangle of A's property on the north, but A was not asked to join in the agreement!

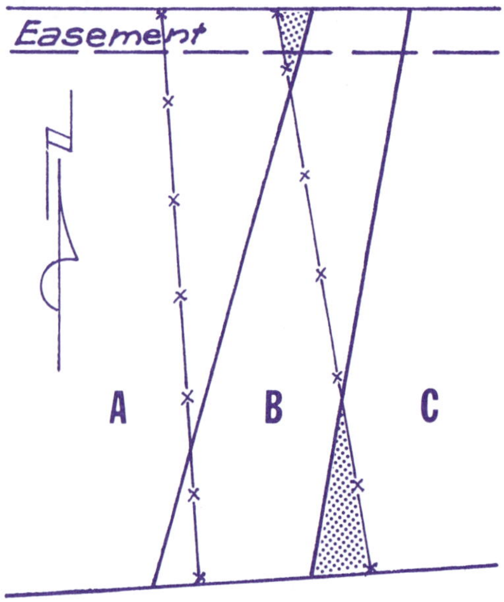

Fig. 66

Now, if a surveyor were asked to survey these lands, he should show *all* these facts and submit to his client that through means of a properly written agreement by an attorney, with words of conveyance, etc., including all parties concerned, the title from the old lines to the occupied lines would be adjusted.

This situation would be interesting for a right of way agent working on the acquisition of an easement along the northerly line of the lands of A, B and C, whether he came upon it before the agreement was made or afterward. In the case of C's grant, if prior to the agreement he used the form, "That portion of C's land included within the easement strip;" he would get no legal rights from C to the portion between C's westerly line and the fence to which C claimed. And if he used the same form after the agreement, he would get no legal rights from C to the portion between B's original westerly line of record and the fence to which C would claim under the agreement because A did not consent to the relinquishment of *his* title to any portion easterly of the fence line. If this agent used the limited form description covering only the segment of the strip included within the respective titles after the agreement, he would have no problem regarding the triangle between the newly agreed-upon fence line and the still-on-record deed line between A and B. The agent's inspection of the ground would lead him to believe that the area immediately east of the fence would be appropriate for a new pipeline easement, and of course in talking with the people on the east, he was led to believe (and without wrong intent) that they would have the right to grant his company that easement. On that basis, he might not even suspect that his acquisition of this particular piece of right of way would be null and void for having a grantor who could not grant. This is an example of how wary the right of way agent must be and how valuable it is for him to get a title attorney's abstract or title company's report on the vesting of the land in which he is interested as well as a survey with which to compare them.

In this case, it would be most certain if the agent could get A, B, and C to all sign one instrument covering the easement from the easterly line of C to the westerly line of A.

As hereinbefore mentioned, the "practical location," that is, the execution and exchange of documents of conveyance by the holders of fee title based on an agreed line, that legally establishes a formerly questionable status is desirable.

Adverse Possession

In the case of adverse possession, however, many conditions must prevail such as possession in actuality, adversely, with continuousness, hostility, openness, notoriousness, claim of right and exclusiveness; color of title under some statues is not always essential[63] but it does carry evidentiary value;[64] in some jurisdictions, the payment of taxes is a prerequisite but is not sufficient in itself; and even though the rules of the particular state are based on statutes of limitations, "In no state is the statute of limitations self-sufficient."[65] Many

[63]Roesch v. Gerst, 138 P. 2d. 846, 18 Wash. 2d. 294.
[64]Lincoln v. Mills, 2 So. 2d. 809, 191 Miss. 512.
[65]Clark on Surveying and Boundaries, Sec. 556.

decisions in different states have attached a variety of interpretations and limitations onto the statutes of adverse possession and its productivity of title.

In early times the possession of land came about by "seizin" which was thought to "hold" the legal rights of the individual. If another took physical (adverse) possession, then the one in "seizin" was disseised and lost his interest in the land. In the beginning, the only remedy to recover the land was for the disseised to oust the disseisor by force promptly before losing the right to regain his "seizin."

Although "adverse possession" and "prescription" are sometimes considered interchangeable, their differentiation is based mainly upon whatever term is correct in the state. California uses both terms, "adverse possession" for an occupant's claim of full ownership and "prescription" for the claims of an adverse use of an easement. Again, though, these rules vary from state to state — some accept prescriptive rights as establishing title to easements while others require adverse possession where the intruder seeks ownership of the land he occupies. In either case the rights are acquired because the true owner, for a continuous, prescribed period, fails to oust the continuing intruder from his land, or stop the intruder from using his land as the site of an easement.

It is not the object here to delve into all the ramificiations of adverse possession but to point up the fact that there is such a variety of rules that you must be very wary of getting involved in any decisions on the matter. In any event, you should digest as much information as possible by researching the statutes and court cases of your area in order to be able to discuss the matter with some comprehension of those rules and some of the problems involved.

Descriptions

If you are asked to write descriptions concerning an agreement line or a line of possession, be fully aware of the *difference* between the lines of title (record) and the lines of occupation shown by survey and then state the facts in no unmistakable terms. Referring back to Figure 65, the two descriptions for the exchange of title might read thusly:

(L to M)

That portion of Lot 4 of the ___ Tract in the City of ________, County of ________________, State of ________________, as per map recorded in Book ___, Page ___, of Maps in the office of the

County Recorder of said County lying easterly of the following described line:

Beginning at the 2″ x 2″ stake at the southeast corner of said Lot 4; thence westerly 19.88 feet along the south line thereof to a 1½″ iron pipe bearing LS tag No. 1234 set for the southerly terminus of an agreement line; thence northerly to a 1½″ iron pipe bearing LS tag No. 1234 set for the northerly terminus of said agreement line on the north line of Lot 5 of said ___ Tract distant along said north line 17.36 feet easterly from a 2″ x 2″ stake at the northwest corner of said Lot 5 as shown on a map attached hereto and made a part hereof.

(M to L)

That portion of Lot 5 of the ___ Tract in the City of ___, County of ___, State of ___, as per map recorded in Book ___, Page ___, of Maps in the office of the County Recorder of said County lying westerly of the following described line:

Beginning at the 2″ x 2″ stake at the southeast corner of Lot 4 of said ___ Tract; thence westerly 19.88 feet along the south line thereof to a 1½″ iron pipe bearing LS tag No. 1234 set for the southerly terminus of an agreement line; thence northerly to a 1½″ iron pipe bearing LS tag No. 1234 set for the northerly terminus of said agreement line on the north line of said Lot 5, distant along said north line 17.36 feet easterly from a 2″ x 2″ stake at the northwest corner of said Lot 5 as shown on a map attached hereto and made a part hereof.

Remembering that the engineer, the surveyor and the scrivener are not permitted to practice law and therefore do not prepare documents, you will confine your expertise to the writing of the description and then submit it to the attorney for incorporation into the proper document.

The encroachment of a building over the line of title is sometimes unknown for many years, perhaps not even until the adjoining building is removed and a survey made in the cleared area reveals that encroachment. The avenues open for settlement of the problem are: (1) An agreement line with proper conveyance of title from the old line to the new one; (2) A court action if all the requirements are present to support it; (3) Execution of a document creating an easement or giving a license for the continued occupation of the building. By whichever procedure is followed, you may be asked to write a description of the concerned area. Again, be explicit by the use of proper words designating the line(s) of title compared with those of occupation.

For an example, consider Figure 67 showing the encroachment of a two-story brick building onto Lot U. Settlement was reached on the basis of the owner of Lot V buying the triangular part encompassed by the building proper and the owner of Lot U giving a 1.35 foot wide underground easement adjoining the building to accommodate the foundation. The two descriptions could be prepared thusly:

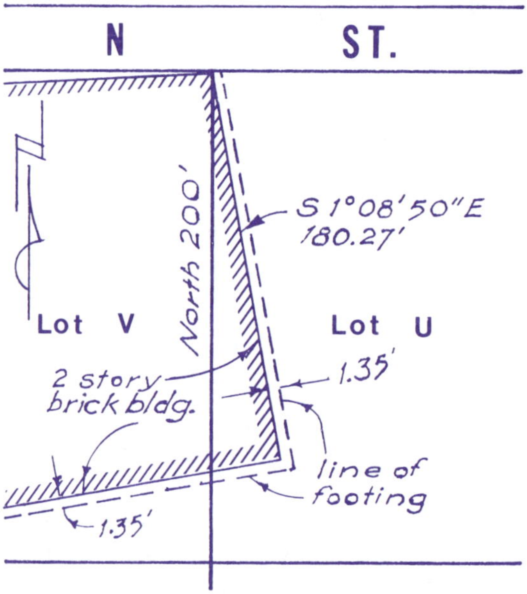

Fig. 67

Parcel 1: That portion of Lot U of the Haymis Industrial Tract in the City of ___, County of ___, State of ___, as per map recorded in Book ___, Page ___, of Maps in the office of the County Recorder of said County described as follows:
Beginning at the northwest corner of said Lot U, being also the northeast corner of an existing two-story brick building; thence South 1° 08′ 50″ East 180.27 feet along the easterly face of said building; thence at right angles along the southerly face of said building South 88° 51′ 10″ West 3.6 feet to the west line of said Lot; thence North to the point of beginning.

Parcel 2: An easement in a strip of land 1.35 feet wide in a portion of Lot U of the Haymis Industrial Tract in the City of ___, County of ___, State of ___, as per map recorded in Book ___, Page ___, of Maps in the office of the County Recorder of said County, the westerly and northerly line of which is described as follows:
Beginning at the northwest corner of said Lot U, being also the northeast corner of an existing two-story brick building; thence South 1° 08′ 50″ East 180.27 feet along the easterly face of said building; thence at right angles along the southerly face of said building South 88° 51′ 10″ West, 3.6 feet to the west line of said lot, the easterly and southerly lines to be lengthened to their common intersection and shortened and extended respectively to meet the north and west lines of said Lot. The upper limit of said easement is the surface of the existing ground and the lower limit of said easement is 3.00 below said surface.

Notice that the stated purpose of the easement, which would be "for foundation purposes," is absent because it is the business of the attorney who draws the documents to create the proper wording for the easement. Of course, you should remind the attorney that this will be needed.

In any subsequent sale of said Lot U, the description of the lot would have to carry an EXCEPTION of the triangle described as above and the remainder of the lot would be SUBJECT TO the easement as described above.

Any time the fee area of ownership is encumbered by an easement of record or physical occupation, the description should be followed by "Subject to . . . . " and describe the areas affected, or omit those areas, either by a perimeter form or by EXCEPTION and then add separate parcels for the affected areas.

PROJECT

The map shown in Figure 68 is of a recent survey. The new owners had a general idea of what their boundaries were supposed to be, but they wanted to know what the field measurements were. They have given you a copy of the deed [a] and map and asked you to write a new description and show all encumbrances.

FACTS: Fresh Grain Corp. acquired this property from Mountanare Development Co. by deed recorded in Book 9824, page 362 of Deeds.

10.12

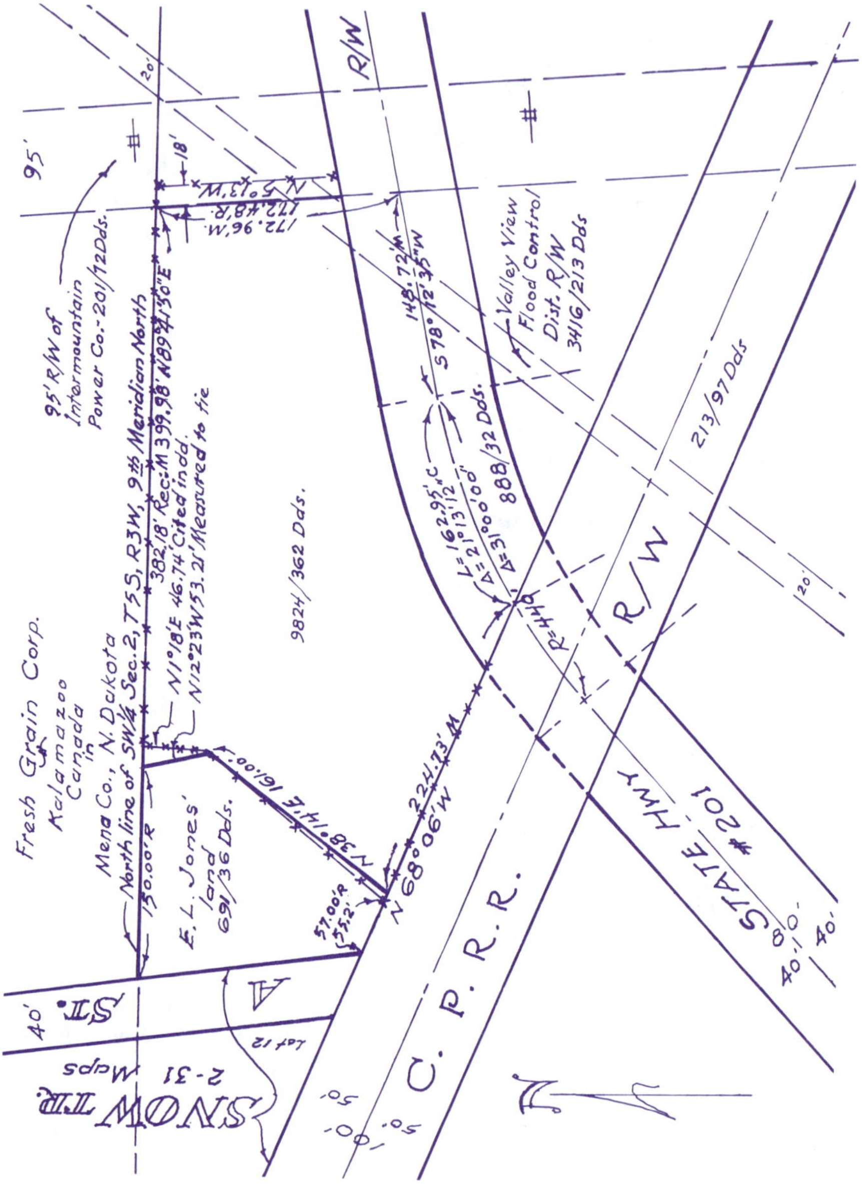

Fresh Grain Corp.
Kalamazoo
Canada
Mena Co., N. Dakota
North line of SW¼ Sec. 2, T5S, R3W, 9th Meridian North
95' R/W of Intermountain Power Co.- 201/72 Dds.
95'
9824/362 Dds.
E. L. Jones' land 691/36 Dds.
Valley View Flood Control Dist. R/W 3416/213 Dds
888/32 Dds.
213/97 Dds
R/W
STATE HWY. #201
C. P. R. R.
SNOW TR.
2-31 Maps
A ST.
40'
Lot 12

Fig. 68

Note a: Description on deed to Fresh Grain Corp. in 9824/362.
Begin at intersection of west line of R/W of Intermountain Power Co. per 201/72 of Deeds with ¢ of State Highway #201 per 888/32 of Deeds; thence along said ¢ westerly to the northerly line of the C.P.R.R. R/W, 100 feet wide; thence westerly along said R.R. to the southeast corner of Jones' land [b] per 691/36 of Deeds; thence along Jones' east line to the north line of the southwest ¼ of Section 2, T5S, R3W, of the 9th Meridian North; thence East 382.18 feet to said west line of Intermountain Power Co. R/W; thence S 5° 13′ E 172.48 feet to the point of beginning.

Note b: Description of Jones' land.
Begin at intersection of north line of southwest ¼ of Section 2, T5S, R3W, of the 9th Meridian North and the easterly line of "A" Street as shown on a map of the SNOW TRACT recorded in Book 2, page 31, of Maps; thence southerly along said "A" Street to the northerly line of the C.P.R.R. R/W as per 213/97 of Deeds; thence S 68° 06′ E 57.00 feet along said R.R.; thence N 38° 14′ E 161.00 feet; thence N 1° 18′ E 46.74 feet to a point on said north line easterly 150.00 feet along said line from the east line of said Street; thence west 150.00 feet to the point of beginning.

ASSIGNMENT

Write a complete (including caption) new description of the Fresh Grain Corp. using and incorporating the survey information shown on the map. At the end show, "SUBJECT TO . . . . . " and describe by literal (not metes and bounds) reference the items and/or areas affecting this land.

Also write a supplemental statement regarding any land areas being used by the owner which are not within the described area.

10.14

PROBLEMS

1. Can taxes paid on questionable property be a basis for claim of title by adverse possession? Explain.
2. Will acquiescence in and to a fence line, that does not agree with the supposed line described in the deed, ripen into an agreement line? Explain.
3. When a surveyor can establish a line described in a deed on the ground and it disagrees with a fence line near it, how is acquiescence in the fence line treated?
4. Explain some differences between "prescription" by common law and "adverse possession" by statutory law.
5. Will an "Agreement" for an agreement line create a perfect title for conveyancing? Explain.
6. In the case of a physical encroachment across the line of ownership, name three methods for settlement of the encroached-upon area.
7. Define a "practical location" of a boundary.

Chapter 11

Writing Descriptions

Framework

The historical background of different land classifications and conveyances in them, the types of records, recordation systems, maps, description fundamentals, including words and phrases, evaluation of controls, applicability of codes, areas, curves, ramifications of boundaries, analysis and interpretation of descriptions, monuments, streets and occupation vs. title, all have been discussed as background preliminary to the study of creating and writing legal descriptions.

It has been said that descriptions which must leave interpretation to "construction" leave some uncertainty.[66] Rules of Construction are needed and used when the wording is not clearly stated. "All rules relating to the comparative dignity of calls are designed to aid in determining which call or calls of a survey were made by mistake."[67] "A good draftsman should avoid placing his client in the position of a supplicant for mercy."[68]

Before starting to write a description, a review of the Do's and Don't's on page A.1 and A.2 of the Appendix will sharpen your memory concerning some of the mechanics involved in such a project.

Parts

A description of the outline of a certain area is the proper grouping of words which delineates one specific piece of land and which cannot apply to any other piece of land. It is called legal because it must stand up under law and survive legal attacks. The parts of a description are essentially the caption, the body and qualifications. Their use is analogous to the use of a camera. You first point it in the general direction of the subject — this is comparable to the caption which embraces the general but limiting area of the description. Next you focus the lens onto the specific subject and take the picture — this is comparable to the body of the description which defines the details of the outline. Qualifications add to, subtract from or encumber the recited details. In some cases, all of the necessary facts can be embraced within the caption alone.

Your slogan should be, "Concise Clarity Without Ambiguity." Do not use time or space for superfluous words, but, on the other hand, do not omit words or phrases that are necessary. Ask yourself the ques-

[66]Article in 38 Cal. Law Review 426.
[67]Gilbert v. Finberg (Tex. Civ. A.), 156 SW 507.
[68]David E. MacEllven, Counsel, Southern Title & Trust Company.

tion, "Will the omission of these words change the meaning?" or "Will the use of these additional words add to or subtract from the actual meaning of the whole phrase?"

Forms

There are several fundamental formats for describing a piece of land; they are by reference to: —

1. The single parcel outlined on a map with a unique identifier assigned to it.
2. A part of a large parcel which can be described by either:
   a. the long form known as metes and bounds, or
   b. the short form expressed as the (south) so many feet of . . . . .
3. Boundary deeds of record.
4. A strip of land taken out of one or more parcels and described either singly as one whole piece or in multiple by separate parcels.
5. Section land or subdivisions thereof described under government nomenclature.

(The last two will be discussed in subsequent chapters.)

The *simplest type* is the use of a lot number in a recorded subdivision as shown on a map; however, there are some other items that belong with this and together they should appear in the following order:

a) Lot or Parcel (number or letter)
b) Block (number or letter)
c) Tract (name or number)
d) City (if any)
e) County (or Parish or Borough)
f) State
g) Map recorded (Book and Page with correct designation for the map book)
h) Public office (Recorder's office, Clerk of Court, Probate Judge, etc.)

For the first format, one of its applications to the facts would be as follows: —

> Lot No. 9 of Block D in the Higginsville Tract in the city of Downsdale, county of Summertown, state of South Dakota, as per map recorded in Book 17, Page 32, of Maps in the office of the County Recorder of said County.

Another simple example using this form is:

> Lot 1 in Block 2 of Tract No. 3 in the city of Kalamazoo, county of Sheridan, state of Oregon, as per map recorded in Book 10, Page 32 of Maps in the office of the County Recorder of said county.

While this is the simplest form of all, it can be easily adapted to other conditions as set forth in the second format. When you have only a portion of a lot to be described, preface the caption with "That portion of" and then follow the caption with "described as follows:"

As in all uses of good English grammar, if there is only one lot, you use the singular form:

> "That portion of Lot 1 in Block 2 . . . . described as follows:"

If there are portions of several lots in the same tract or several sections in the same township, you use the plural forms, such as:

> "Those portions of Lots 6, 7, 8 and 9 in Block 3 . . . . described as follows:"

or

> "Those portions of Sections 10, 11 and 12 in Township 3 South, Range 6 West . . . . described as follows:"

For combining parcels out of different tracts where you must recite each lot and each tract with its own recorded reference, use the form:

> "That portion of Lot 3 of the Smith Tract in the city of Hamburg, county of Newton, state of Oregon, as per map recorded in book 56, page 32 of Maps, records of said county, and that portion of Lot 20 of the Oldham Tract in said city as per map recorded in book 72, page 21 of Maps of said records, described as a whole as follows:"

In each case, the metes and bounds will be added.

Another suggestion for the first part of the second format would be:

> That portion of Farm Lot 114 in Tom's Vineyard Tract in the county of Wild Life, state of Wyoming, as per map recorded in Book 76, Page 28, of Maps in the office of the County Recorder of said County, described as follows: . . . . .

Describing a part of the above could be as follows:

> The South (or north, etc.) 250.00 feet of Farm Lot 114 in Tom's Vineyard Tract in the county of Wild Life, state of Wyoming, as per map recorded in Book 76, Page 28, of Maps in the office of the County Recorder of said County.

or:

> The west 196.00 feet of that portion of Farm Lot 114 in Tom's Vineyard Tract in the county of Wild Life, state of Wyoming, as

per map recorded in Book 76, Page 28, of Maps in the office of the County Recorder of said County, described as follows: Beginning at the Southwest corner of said Lot; thence N 89° 36′ E 452.00 feet along the south line of said lot; thence N 28° 24′ E 238.79 feet; thence N 12° 54′ W 182.36 feet; thence N 75° 52′ W 550 feet more or less to the west line of said Lot; thence S 0° 24′ W to the point of beginning.

Also instead of continuing courses following the word "thence," a program of numbered courses may be used thusly:

That portion of Farm Lot 114 in Tom's Vineyard Tract . . . . . . of said County, Beginning at the Southwest corner of said Lot; thence the following courses:
1 — N 89° 36′ E 452.00 feet along the south line of said lot;
2 — N 28° 24′ E 238.79 feet;
3 — N 12° 54′ W 182.36 feet;
4 — N 75° 52′ W 550 feet more or less to the W line of said lot;
5 — S 0° 24′ W to the POB.

The last two above forms illustrate the "free" line description, that is aside from the lot boundary, the other lines are free from any joinder parties. The opposite form using only boundary references would be expressed thusly: (See Figure 69)

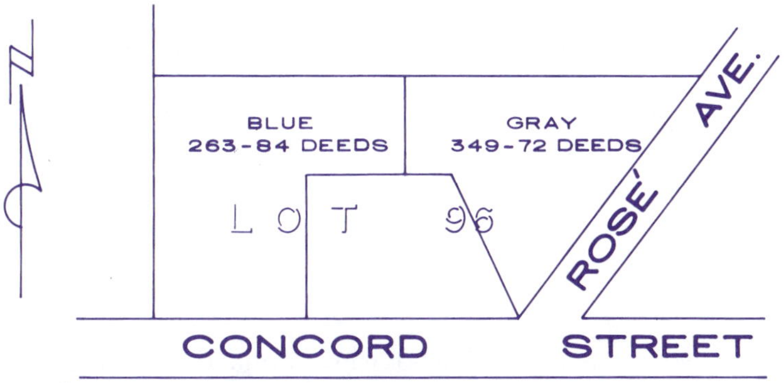

Fig. 69

That portion of Farm Lot 96 in Tom's Vineyard Tract . . . . . etc. bounded as follows:

On the west and north by the land of Raymond Blue as described in deed recorded Feb. 21, 1927 in Book 263, Page 84 of Deeds.

On the north and east by the land of Tom Gray as described in deed recorded Jan. 14, 1929 in Book 349, Page 72, of Deeds.

Note that it is *unnecessary* to state, "On the south by the south line of the Lot." or "On the south by the north line of Concord St." These are facts shown on the map described in the caption and which therefore includes all facts shown thereon.

Although the above type of description using boundary deeds of record is perfectly good for title, it is difficult, not only for the surveyor in the field but also for the draftsman, to use without copies of all the reference material. It is much more desirable to describe the lines as given in the deeds and then correlate them with the deed references.

There is still another type of short form description and it is used in lieu of metes and bounds. This can be adapted and used with any of the above cited caption forms, thus:

"The north 40 feet of Lot 4 . . . ." etc. (Figure 70)

or

"The east 50 feet of the south 80 feet of Lot 2 . . . ." etc. (Figure 71)

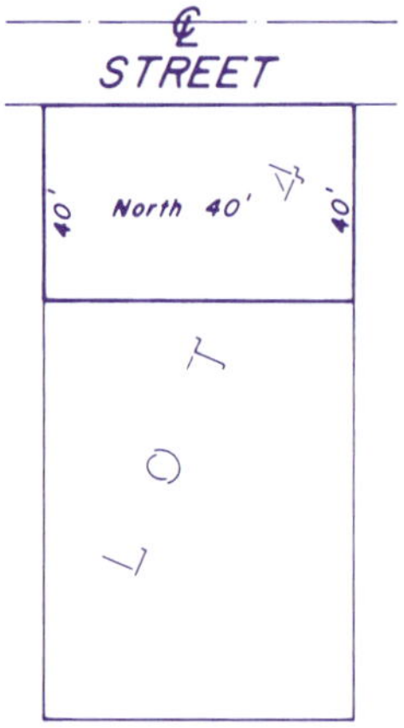

Fig. 70

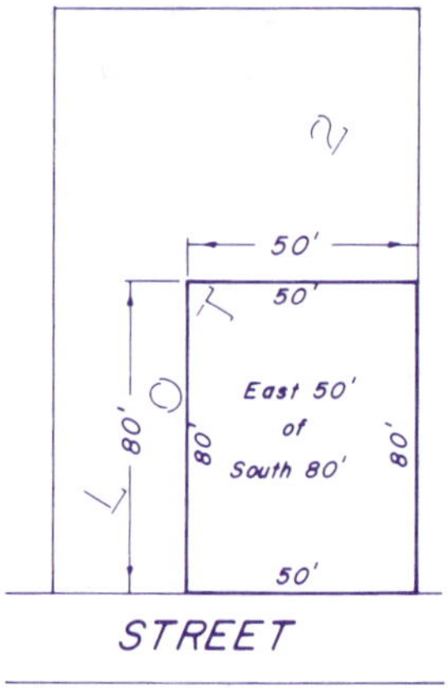

Fig. 71

"The west 3 acres of the southwest quarter of the southwest quarter of the southwest quarter of Section 3. . . ." etc. (Figure 72)

or

"The south 8 acres of the southeast quarter of the southwest quarter of the northwest quarter of Section 12. . . ." etc. (Figure 73)

You will notice that in these forms, all the information is included in the caption. Variations of this principle and limits of applicability have been expanded upon in Chapter 4.

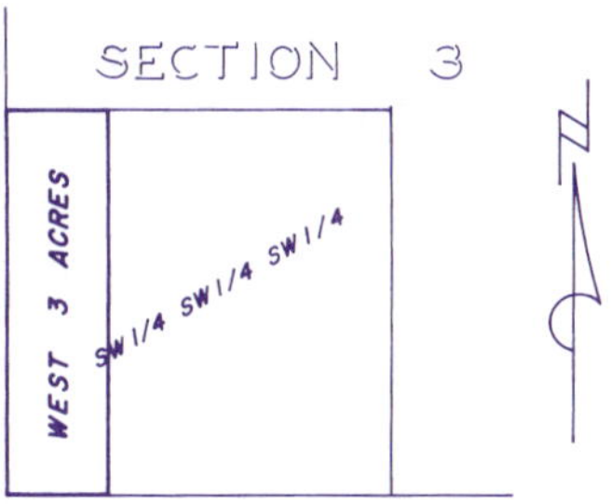

Fig. 72

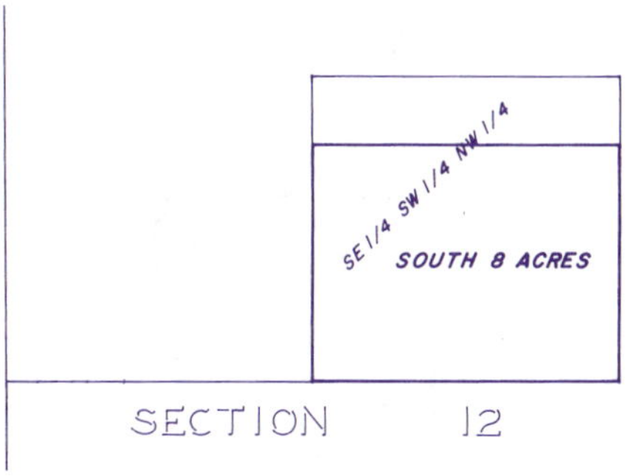

Fig. 73

A word about the phrase, "fractional lot," which is sometimes used to denote a cut-up lot. No matter how many parts are taken out of a lot shown on a map used for reference in the conveyancing of those parts, the lot itself is still a "whole" lot as shown on that map and it should not be referred to as a "fractional" lot.

Exceptions

A different application of the short form description is by the use of exceptions. This is often simpler and safer than by metes and bounds. The entire caption is used as before and then an exception follows it, such as:

"EXCEPT the south 30 feet thereof." (Figure 74)

or

"EXCEPT the south 10 feet of the north 30 feet thereof." (Figure 75)

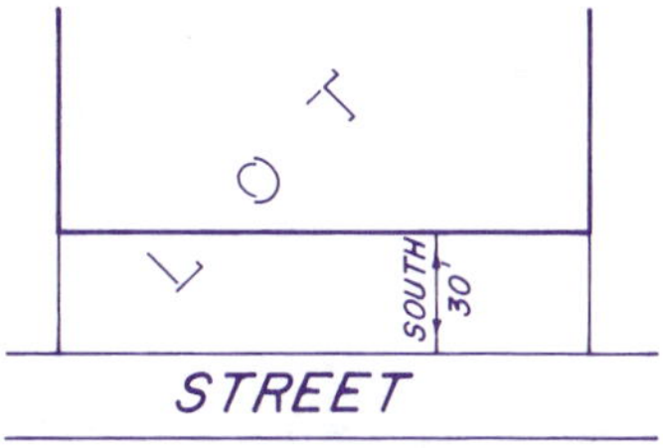

Fig. 74

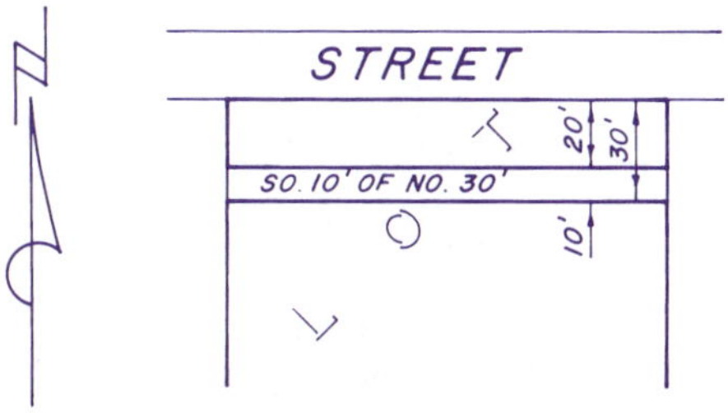

Fig. 75

Be sure that you do not put the exception, or exceptions, in the caption because it creates confusion. If you have

"The West 200 feet except the West 30 feet of Parcel 2 as per map filed in . . . .(etc.) . . . . EXCEPT the East 100 feet." (Figure 76)

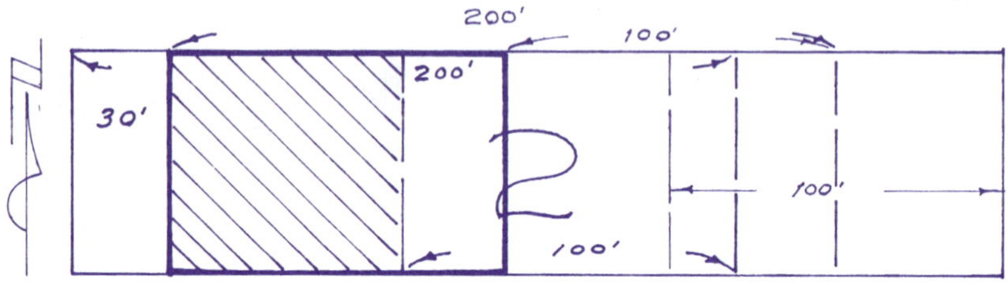

Fig. 76

you do not know whether the "West 200 feet" is to be excepted before or after the exception of the "West 30 feet" and, furthermore, you do not know whether the exception following the description is to be taken before or after the one in the caption. These conditions are ambiguous and the conveyance of title is indeterminate.

Obversely, it would be acceptable to express a double exception as follows:

"Parcel 5 as per map recorded in . . . .(etc.) . . . .
EXCEPT the East 30 feet.
ALSO EXCEPT the South 15 feet." (Figure 77)

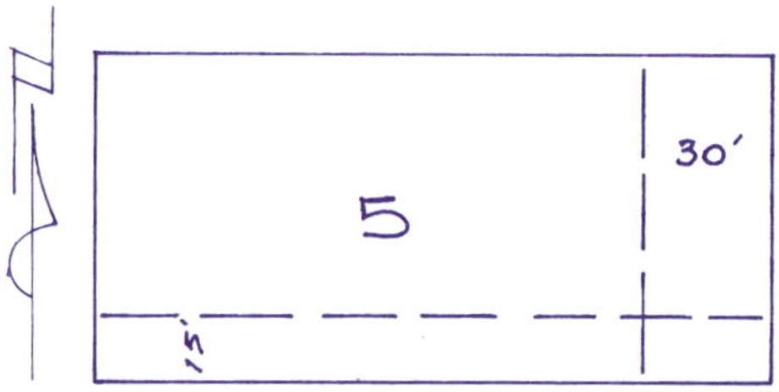

Fig. 77

Two factors make this form satisfactory: 1) the exceptions **follow** the caption; 2) the second exception is preceded by the word "ALSO" which makes it independent from the first exception. With any multiple number of exceptions, **all** but the first one should be preceded by the word "ALSO."

A seldom used but desirable type of description is "That portion of" plus the entire caption, plus "lying west (east) of the following described line."

This is a very safe way because you describe only one line and each part lies on its side of that one line; you can have no conflict by this method. For example:

> "That portion of Lot 16 in the Sunnyvale Tract in the county of Willowbrook, state of New Mexico, as per map filed in Book 72, Page 31, Records of said county, lying easterly of the following described line:
>
> Beginning at a point on the south line of said lot West 36.84 feet from the southeast corner of said lot; thence North 8° 23′ West 48.27 feet; thence North 2° 47′ East 53.06 feet to the north line of said lot."

Sufficiency

The value of the description depends upon its being sufficient and unambiguous. Even a county omitted from a description has been held to render a trust deed and the subsequent trustee's sale void.[69] When performance of a contract depends upon the description, it must contain all the necessary requisites so that parol evidence is *not* needed in order to correctly identify the property; otherwise, the contract may be void for lack of being able to properly locate the land.

If a description is sufficient for a competent surveyor to locate the land on the ground, with or without extrinsic evidence, it is considered sufficient as between the parties.[70] If discrepancies are discovered, corrections may be made, either through court action or corrective deeds. Parol or extrinsic evidence is not admissible in court where it adds to or varies a definite call. There have been inroads made in the parol evidence rule but not enough for relaxation of the surveyor's rule. Your description should be so clear that the land you mean to cover is unmistakable. If so, attempts to vary your meaning will never succeed.

Beginnings

For metes and bounds descriptions, a competent and verifiable point must be either in existence, or properly established in relation to an acceptable point, or points, for the beginning and control of the land described from it. The point may be represented by a post, a tree, an iron pipe, or rod, an intersection of two streets or the intersection of two lines having dissimilar qualifications, etc., etc. but it must be clearly identified or its value will be lost. Furthermore, the designa-

[69]Saterstrom v. Glick Bros. Sash, Door & Mill Co., 118 CA 379, 5 P 2d. 31.
[70]Blume v. MacGregor, 64 CA 2d. 244, 251, 148 P 2d. 656.

tion of the point of beginning must not only cover its physical attributes but should include references to its legal relationship to other matters. For instance, "Beginning at a 3″ x 3″ hickory post at the northwest corner of Boman's 10 acre tract as shown on the map thereof recorded in Book 10, Page 4, of Miscellaneous Maps in the office of Public Records," gives both the physical description and legal reference satisfactory for control.

When the true point of beginning of the subject property is not yet established, it must depend upon a remote point of beginning which is already recognized. The transition between the two is accomplished through several different combinations of words. The first part (from the recognized point) usually starts with just "Beginning at" and after going through the necessary courses and distances, arrives at the "True Point of Beginning," after which the subject property is described and the closing course must then return "to the True Point of Beginning."

Some scriveners prefer to start the first part with "Commencing at" and go to the "Point of Beginning" and return to the "Point of Beginning."

There is no rule concerning this and the combination of words is immaterial so long as the *distinction* between the two types of points is made clear.

Occasionally, the two separate points of beginning can be combined as follows:

> "Beginning at a point in said Lot N 41° 28′ 10″ E 94.68 feet from the southwest corner of said lot; thence . . . . ."

or

> "Beginning at a point N 82° 16′ W 252.16 feet from a point which is S 26° 34′ W 362.78 feet from the northeast corner of said Section 12; thence . . . . ."

It is not recommended that more than one, or, at the most two, course and distance recitals be used in this form.

Base of Bearings

This is written with the first word spelled as either *base* or *basis.* Webster makes no differentiation between their meanings as related to our use in descriptions, while Black adds nothing to the interpretation, so either way is acceptable.

Bearing is an angular quantification which expresses the proceeding direction of a line; in other words, the direction in which the line is proceeding as we generate it or retrace it. On a map, the same angular direction of a line may be either way; for example, "N 78° 52′ 34″ E" or "S 78° 52′ 34″ W," but in a written description it is positively necessary that the direction of the angular quantity be expressed with consecutive continuity in the direction of travel from beginning to end. In other words, to copy bearings from a map which does not express continuing directional control of a boundary is unacceptable. A line's bearing is comparable to a line's dimension in that each one in its particular realm establishes the limitation of that line.

The establishment of a *base* from which bearings in the description can evolve is developed from such items as survey lines, boundary lines in or derived from a deed, lines shown on a record map, pairs of coordinates, pairs of physical monuments or an angle to a line turned from an astronomical observation.

It is advisable that following the "Point of Beginning," there should be established, by recitation of a bearing, the basis upon which subsequent bearings will depend for orientation. In the absence of such a statement, if there is reference to a map in the caption or in the forepart of the body of the description, then the assumption can properly be made that any one or more of the bearings shown on the map are acceptable for control.

There may be more than one base of bearings in a description by reason of the fact that the adjoining land to which ties are made carries a different base of bearings and therefore your angular alignment must be adjusted to conform. Examples of this type of situation are cited in Chapters 4 and 5 supra. The fewer changes of bearing basis that are in a description, the easier it will be to decipher and understand. The more desirable way is to have sufficient survey information to permit the continuity of angular relationship, together with the recitation of external controls.

Ties and Boundaries

Discussion has been had concerning adjoining properties and their value as limiters of title. This, in essence, is the reason for including "ties" in a description so as to bring lines measured on the ground into agreement with the lines of ownership on record.

Ties are cited in several ways such as to physical monuments, lot corners, tract boundaries, street intersections or side lines, right-of-ways of railroads, canals or highways, neighboring property, buildings or walls, etc. Sometimes, however, the thing tied to is faulty. From investigation, this may be found to be a wrongly referred to or misplaced monument, a rebuilt wall not in the same place as the original one to which the description referred, a deed incorrectly written which makes it void as a boundary or a survey which proves to be erroneous insofar as the line of title is concerned.

The point to remember here is that when you are reading descriptions, be sure that the things to which ties are made exist with validity. Likewise, when you are creating descriptions and incorporate references to boundary deeds, monuments, etc., be certain that each item to which you tie is incontrovertible.

Sometimes a tie is made to a point which must hold a certain distance (or distance and bearing) from another point. Consider the situation shown in Figure 78 with the description worded thusly:

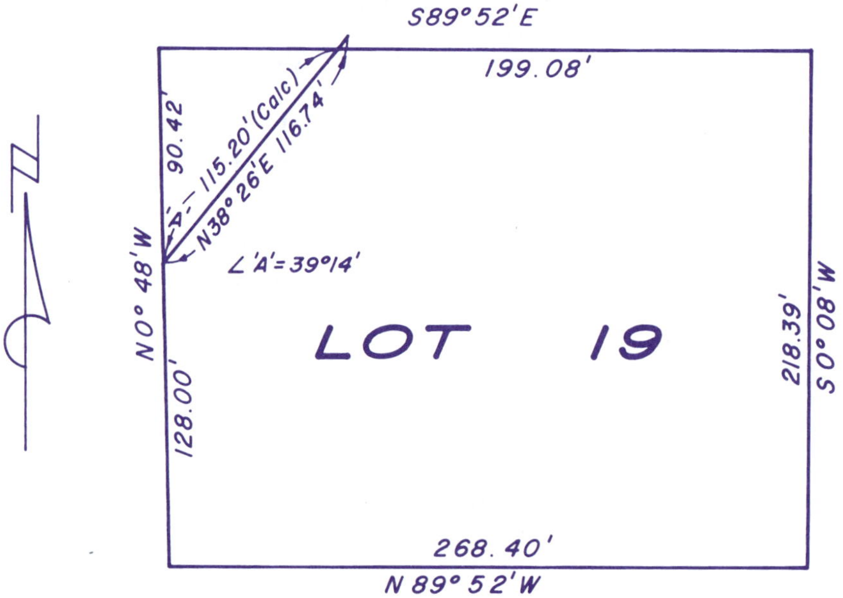

Fig. 78

"That portion of Lot 19 in Tract . . . . (etc.) described as follows: Beginning at the southwest corner of said Lot; thence along the west line thereof N 0° 48′ W 128.00 feet to a point on said line S 0° 48′ E 90.42 feet from the northwest corner of said Lot; thence on a deflection angle of 39° 14′ to the east, N 38° 26′ E 116.74 feet to the north line of said Lot; thence along said line S 89° 52′ E 199.08 feet to the northeast corner of said Lot; thence S 0° 08′ W 218.39 feet to the southeast corner of said Lot; thence N 89° 52′ W 268.40 feet to the point of beginning.

The initial distance of 128.00 feet must yield to the tie of 90.42 feet from the northwest corner of the lot. The next course must hold the angular relation from the west line and stop at the north line, even though the distance given is too long. Each of the succeeding distances will also have to yield to the true corner of the lot.

A much shorter way would be the "dividing line" form as follows:

> Lot 19 in Tract. . . . (etc.)
>
> EXCEPT that portion thereof lying northwesterly of the following described line:
>
> Beginning at a point on the west line of said Lot S 0° 48′ E 90.42 feet from the northwest corner of said Lot; thence N 38° 26′ E 115.20 feet to the north line of said Lot.

On occasion, combinations of ties to physical things are made in such a way as to create complicity similar to the problem of the multiple fixed curve situation exampled in Chapter 4 supra. The following description taken from a recorded document illustrates the problem.

> "Beginning at a point on the south line of said northwest quarter distant South 89° 58′ 30″ East 984.12 feet from the southwest corner of the northwest quarter of said Section 11; thence North 2° 48′ 30″ East 431.52 feet to an iron axle on a line which bears North 89° 54′ 00″ East from a point on the west line of the northwest quarter of said Section 11, which is North 2° 49′ 30″ East 429.36 feet from the southwest corner of the northwest quarter of said Section 11; thence North 89° 54′ 00″ East 465.59 feet to an iron axle set North 2° 49′ 30″ East from a point on the South line of the northwest quarter of said Section 11, distant South 89° 58′ 30″ East 1449.83 feet from the southwest corner of the northwest quarter of said Section 11; thence South 2° 49′ 30″ West 432.53 feet to said point on the South line of the northwest quarter of said Section 11; thence North 89° 58′ 30″ West along said South line of the northwest quarter of Section Eleven 465.71 feet to the point of beginning."

Again the question arises as to which call is to hold, the iron axle or the courses cited, "North 89° 54′ 00″ East from a point on the west line of the northwest quarter of said Section 11, which is North 2° 49′ 30″ East 429.36 feet from the southwest corner of the northwest quarter of said Section 11."?

If the "iron axle," being a cited physical monument, is to hold, as is the normal rule of construction, then why give a specific tie to or from another corner? Following this is another tie "to an iron axle" which is also cited with courses from the same quarter section corner. Since no two field surveys are likely to agree with these measurements given, and if the axles were inadvertently set out of this position, which will hold? In addition, it is interesting to note that all this is shown on a Filed Map with more contrasting distances shown to the same points. So the question is, why not merely refer to the Parcel shown on that map instead of writing the metes and bounds description? since all facts shown on the map are included.

Another example of multiple ties are found in a deed which states, ". . . .; thence with Weeks' line, the south line of the NW¼ of Section 3 and a fence line. . . . . ." Are all three conditions actually coincident? Did the deed to Weeks call for the fence line or the quarter-section line? Was the fence line set correctly on the quarter-section line?

Referring back to the problem of the iron pipe set on the R.R.R/W (Figure 46 in Chapter 7), the multiple ties were made with good intentions but the iron pipe reference proved to be wrong. Always reflect on that particular statement held by the courts, "facts that a *correct* survey will show." Multiple ties are acceptable within limits when their certainty is verifiable.

Briefly reviewing the matter of monuments as ties, if they were set at the time of the creation of the deed, and it is so proven, they may control over bearings and distances[71] while, in other cases where there are monuments but they are not recited in the description, they may not hold, especially if they would alter the "designation of course and distances of boundary lines contained in title papers under which both parties derive title."[72]

Using a combination of monuments and boundary deeds as ties, consider the facts shown on the map in Figure 79 for writing a description of Parcel Y. None of the parcels X, Y or Z are on record so they cannot be used as reference for the instant property.

A good way to organize the writing of a description, after setting up the caption, is to list the calls that will control which, in this case, will be:

North 85° East 210.00′ along S'ly line of Lot.
North 14° 30′ East.
Southerly line of DIG Co.
North 83° 20′ East 160.00′.

[71]Potter v. Bonner, 174 NC 20, 93 SE 370.
[72]Trimmer v. Martin, 141 Va. 252, 126 SE 217; Stark v. Adams (Tex. Civil A.), 183 SW 58.

South 70° East.
North 5° West 162.00′ from N'ly line of 2nd. St.
Deed to City recorded in book 182, page 10, OR.
Lot line.

There is no rule that requires you to write every description in a clockwise direction, but in analyzing the fact that there is no distance given for the southerly line of Parcel Y nor for its northeasterly line, the conclusion is that the southerly line should be the closing course going west so in this case that establishes the direction for writing the description.

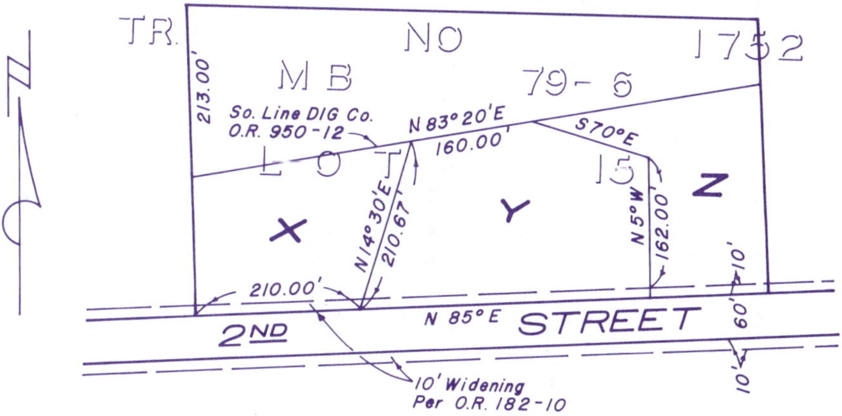

Fig. 79

Now write the caption:

That portion of Lot 15, Tract No. 1752 in the city of ________________, county of ________________, state of ________________, as per map recorded in book 79, page 6 of Maps in the office of the County Recorder of said county described as follows:

Next proceed with the body of the description incorporating the ties previously listed:

Beginning at a point on the southerly line of said Lot North 85° East 210.00 feet from the southwest corner of said lot; thence North 14° 30′ East 210.67 feet to the southerly line of the land described in deed to DIG Co. recorded in book 950, page 12, Official Records.

You could end this tie here and proceed because the boundary would be good for title, but absent the map, you would be in a quandary as to its relationship to the west and north line of the lot so it is desirable to describe the line as it is stated in the deed:

> which line bears North 83° 20′ East from a point on the westerly line of said Lot 213.00 feet southerly along said line from the northwest corner of said lot; thence North 83° 20′ East 160.00 feet along said southerly line; thence South 70° East to a point on a line North 5° West 162.00 feet from the northerly line of 2nd. St. as widened per deed to said City recorded in book 182, page 10, Official Records; thence South 5° East 172.00 feet to said southerly line of the Lot; thence South 85° West to the point of beginning.

To those who would argue or question the validity of establishing the position of the easterly terminus of the line bearing "S 70° E," consider the geometry of the facts. Since the next line at that easterly terminus is perpendicular to 2nd. St., the intersection can be determined by establishing a line parallel with 2nd. St. and 162.00 feet northerly from the 10-foot widening line (or 172.00 feet from the original lot line) and the bearing of the line S 70° E will intersect at only one point. Of course, the mathematical derivation can be obtained from multiple closures and line-bearing intersects. Likewise, the missing distances can be obtained from a field survey. Here is another example of being able to write a description which is perfectly good for title without all the measurements.

In those rare cases where you find a gap between two properties because each one was described from opposite control points and later a survey revealed an excess of measurement on the ground, be careful how you handle it. There are three possible answers to the situation shown in Figure 80. If Parcel B was sold several years after the sale of Parcel A, it has been held in some cases that B was entitled to the gap because the former owner of Lot 6 had no inten-

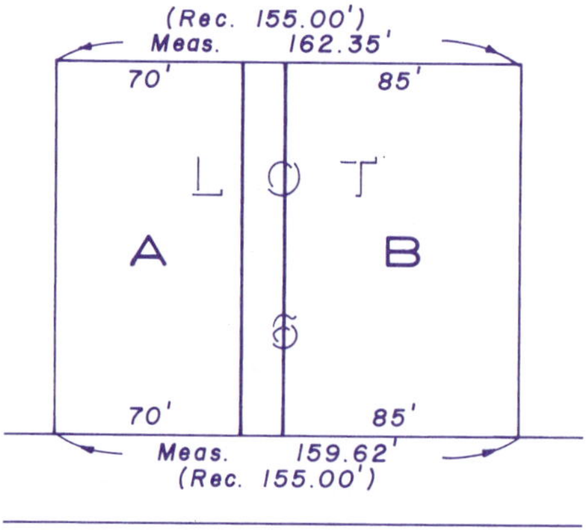

Fig. 80

tion of retaining any part of it. If Parcels A and B were sold simultaneously, or within a few months of each other, it has been held in some cases that the gap should be prorated between them, also on the theory that the common owner never meant to keep any of the lot from his sale but intended to divest his interest on the proportionate basis of the two deeds. It has also been upheld that the gap was not disposable in any way other than by direction of the grantor of Parcels A and B.

Conversely, if there is an overlap, both principles, that of the second buyer receiving whatever is left, and the proration of shortage in the case of simultaneous deeds, could apply. In each result, an enforcement of the "intent" of the grantor and the justifiable expectations of the grantee is the aim of the court.

Consequently, if you are asked to write a description for all or part of a parcel bounding a gap, do not go into nor include any part of that gap until or unless your client has a deed on record for it, either from the legal owner or by court decree or if an attorney or title company will vouch for it or insure it. Likewise, in case of a shortage, back off from it until there is a legal settlement regarding the boundary.

Let us review the effect of ties and boundaries on courses and distances in the sense of priority of calls. In Chapter 4, "Distances on boundary, without ties to any of the foregoing." was placed ahead of "Bearings on boundary or angle between lines, without ties to the foregoing," subject to interchangeability dependent upon circumstances. In the citation of comparative dignity of calls given by different courts in their consideration of rules of construction, courses are given precedence over distance usually on the basis that the distance is tied to some thing and must therefore yield to the ground position of that thing. The course, however, must also give when tied to a thing; therefore, when a line is subject to the control of a bound, both the course and distance must yield. Another aspect of distance yielding to course is cited in 9 C.J. p. 227.[73] "If the courses and distances thus run do not close the survey, it must be accomplished by running the same courses and either lengthening or shortening the distance as each case may require, and in proportion to the length of each line, as called for in the plat and certificate of each survey."[74] This application is sometimes referred to as "the rubber tape" method

[73]Combs v. Valentine, 144 Ky. 184, 137 SW 1080; Pearson v. Baker, 4 Dana (Ky.) 321.
[74]Beckley v. Bryan, 1 Ky. (Sneed) 91.

based on the theory that the length of tape used by one surveyor did not match the same marked length of tape used by the other surveyor. In other words, one or the other, or both, tapes were not standardized. Concerning course and distance, 9 C.J. further states, "The true rule is that one or the other shall be preferred according to the manifest intention of the parties and the circumstances of the case."[75]

The application of the distribution of distances in a series of courses to correct an error of closure, while the angular relation between the segments within the problem area is held, has been a standard method of correction and adjustment for nonriparian meander lines and grant boundaries as set forth in the Instruction for the Restoration of Lost or Obliterated Corners by the United States Department of Interior, Bureau of Land Management. (Sections 5-43 and 5-44 in the Manual of Surveying Instructions, 1973.) This could be one of the reasons why courts have leaned toward subordinating distance to course.

The result of using the more sophisticated survey instruments and electronic distance measuring devices will continue to diminish the differential between angle and distance measurements until perhaps eventually the courts will consider them as an integral unit with no preference.

The application of holding a distance on the boundary without ties prevails when the description cites a flat call such as "N 32° 16′ 32″ W 346.79 feet" with no qualification attached; this means that the line goes exactly that distance. Likewise, if the bearing on the boundary as cited above has no other control, it will hold as is. "Course and distance from a given point is a certain description in itself . . . . ." according to Den V Graham, 18 N.C. 76, 27 Am D 226 (per Ruffin, J.) The bearing and distance combination, as with all other calls considered for comparative dignity, yield in proper obeisance to the circumstances at hand.

Sometimes a passing or incidental tie is made in the description (derived usually from the field notes of the survey) which proves to be the lifesaver. It is most likely to be so when the referred-to monuments such as trees or hedges or walls have been destroyed or rotted away and the incidental tie is the only remaining physical evidence. A substantive example of this is given in a paper delivered by

[75]Loring v. Norton, 8 Me. 61; Green v. Pennington, 105 Va. 801, 54 SE 877; Ruffner's Heirs v. Hill, 31 W. Va., 428, 7 SE 13.

the Honorable Conan Cantwell on LOCATING METES AND BOUNDS[76] wherein he cites the description under consideration in the case of Williams v. Sinclair Prairie Oil Co.[77] as follows:

> "A part of the John Ruddle H. R. Survey, and located about nine miles west of Longview, Texas, and ¼ mile S. of Camp Switch, Texas, and
>
> BEGINNING: 1440 vrs. South of the NWC. of said survey;
>
> THENCE E 570 varas stake for corner from which a pine brs S 72½ deg. E. 38 vrs.;
>
> THENCE N 460 vrs. a stake from which a hickory brs S 35 deg. E 2 vrs.;
>
> THENCE W 570 vrs. a stake from which a pine brs N 75 deg. E 11.8 vrs.;
>
> THENCE South 70 vrs. past the NEC of a 37 acre tract made for W. L. Welborne and continuing on South to the place of beginning and containing 46.2 acres, more or less."

The trees called for at the Northwest corner of the Ruddle Survey had disappeared and the corner was unmarked. Likewise, the trees called for at the other corners of the 46.2 acre tract had disappeared but the Northeast corner of the W. L. Welborne Survey called for was a well recognized and established corner. Since the description called to run from the Northwest corner of the 46-acre tract 70 vrs. "past" the Northeast corner of the Welborne Survey, the Court held that the 46-acre tract should be located by that call.

This decision was also based on another important rule which has been stated by the courts, the rule that where the natural and artificial objects of the grant cannot be identified upon the ground, the proper method of locating the lines and corners will be by course and distance from the nearest recognized and established corner or artificial object with which the field notes are connected.[78]

Reference Material and "Notice"

The many references used in writing legal descriptions can be categorized generally as physical (natural or artificial) or record (public or quasi-public). Accuracy in stating fact is a most pertinent

[76]Sixth Annual Texas Surveyors Association Short Course.

[77]247 SW 2d 422.

[78]Kirby Lumber Co. v. Adams, 127 Tex. 376, 93 SW 2d. 382, Taylor v. Higgins Oil & Fuel Co., 2 SW 2d. 288 and Williams v. Sinclair Prairie Oil Co. supra.

matter in legal descriptions and for this reason, one must not blindly use references but consider them in the light of legal acceptability. This brings into the picture the matter of "Notice" as applied to those references incorporated into descriptions. From Black's L. D. we find it means "knowledge of the existence of a fact or state of affairs; the means of knowledge." It further states that "Knowledge of facts which would naturally lead an honest and prudent person to make inquiry constitutes 'notice' of everything which such inquiry pursued in good faith would disclose."[79]

Continuing, "Notice is either (1) statutory, i.e., made so by legislative enactment; (2) actual, which brings the knowledge of a fact directly home to the party; or (3) constructive. Constructive notice may be subdivided into: (a) Where there exists actual notice of matter, to which equity has added constructive notice of facts, which an inquiry after such matter would have elicited; and (b) where there has been a designed abstinence from inquiry for the very purpose of escaping notice. Wharton." And, "Constructive notice is information or knowledge of a fact imputed by law to a person (although he may not actually have it), because he could have discovered the fact by proper diligence, and his situation was such as to cast upon him the duty of inquiring into it."[80] That is, the law casts on him a "disability to complain" unless he inquired into it.

"Actual notice has been defined as notice expressly and actually given, and brought home to the party directly.[81] The term 'actual notice,' however, is generally given a wider meaning as embracing two classes, express and implied; the former includes all knowledge of a degree above that which depends upon collateral inference, or which imposes upon the party the further duty of inquiry; the latter imputes knowledge. In this sense, actual notice is such notice as is positively proved to have been given a party directly and personally, or such as he is presumed to have received personally because the evidence within his knowledge was sufficient to put him upon inquiry."[82]

" 'Constructive notice' is a presumption of law, making it impossi-

[79]Twitchell v. Nelson, 131 Minn. 375, 155 NW 621.

[80]Baltimore v. Whittington, 78 Md. 231, 27 A 984, Acer v. Westcott, 46 NY 384, 7 Am. Rep. 355.

[81]Jordan v. Pollock, 14 Ga. 145; McCray v. Clar, 82 Pa. 457; Morey v. Milliken, 86 Me. 464, 30 A 102, 105.

[82]Picklesimer v. Smith, 164 Ga. 600, 139 SE 72; White v. Fisher, 77 Ind. 65, 40 Am. Rep. 287.

ble for one to deny the matter concerning which notice is given, while 'implied notice' is a presumption of fact, relating to what one can learn by reasonable inquiry, and arises from actual notice of circumstances, and not from constructive notice.[83] Or as otherwise defined, implied notice may be said to exist where the fact in question lies open to the knowledge of the party, so that the exercise of reasonable observation and watchfulness would not fail to apprise him of it, although no one has told him of it in so many words."[84]

The reason for keeping in mind this matter of notice is that by referring to those matters which carry the weight of legal recognition, as sources of constructive or recorded notice, your descriptions will be more self-sufficient.

Incidentally where two or more maps are recorded covering the same area but each with a different design of subdivision, any one of the map references can be used for a legal description because it is still an official record. Any question concerning the difference of positions of interior lines between one map and another of the same area would depend upon two conditions: 1-the chronological order of any lands granted therefrom; 2-any possible conflict of boundaries dependent upon such grants.

Other than subdivision maps, many maps which are filed in the recorder's office do not meet the requirements of constructive notice and would therefore not be found in a chain of title applied to a search on an owner's property.

Another type of map used for reference in legal descriptions is not in the recorder's office at all — it is the Government Township Plat recorded by the Department of Interior as provided for by Section 751 of Title 43 of the United States Code and filed in the District Land Office. This map also does not come within the purview of constructive notice but when the Patent to the individual is recorded in the official public depository of records, the person reading the Patent will have actual notice that such a map is involved and may thus have a "disability to complain" legally unless he actually looks at the Patent and the Government Plat described therein.

In the case of Stearns v. Title Insurance and Trust Company,[85] one of the questions centered on the use of, and reference to, a filed Record of Survey map, its relationship to the Official Township Plat and

[83]Charles v. Roxana Petroleum Corporation, C.C.A. Okl., 282 F. 983.
[84]City of Philadelphia v. Smith, Pa., 16 A. 493.
[85]18 C.A. 3d. 162.

whether it imparted constructive notice in the same way as a recorded deed would do so.

When a document such as a deed, lease, etc., includes reference to a map in its legal description and is recorded and indexed, it then establishes and carries forward the included map reference into the chain of title insofar *as it applies to its specific part in the description.* For example, if a part of section line shown on a Government Plat is one part of the boundary being described and a part of a recorded tract map is another part of that boundary, each map is applicable insofar as it pertains to its respective part.

Although the court held that the Official Government Surveys filed in the District Land Office and the Record of Survey maps filed in the recorder's office were and do constitute public records, they are not by definition public records *which impart constructive notice.* To impart constructive notice, the "conveyance" must be within the chain of title and these maps were not.[86]

The above finding by the court certainly does not negate the status of those maps as evidence of survey facts regarding boundaries. It is one thing to search a chain of title to establish ownership by legally defined constructive notice, and quite another thing to recover all available information through the avenues of actual notice, both on and off record, to determine boundaries.

In Wisconsin, Illinois and Michigan, constructive notice does not include a name in the chain of title necessarily; they follow a Land Parcel Index which by their definition is constructive notice.

Public and Quasi-Public Records

Generally speaking, those matters found in a public depository which have been placed there for legal recordation thereby giving public notice are by law unalterable. These are classified as public records, and because of that permanency, and because they are indexed and reproduced in the manner defined by the recording statutes, are used for bona fide reference material. Some recorder's offices segregate their material into just a few categories, while others file or record under many subject headings, some of which conform to the statutory rules of constructive notice and others do not.

The ownership of land is based, and carried forward, on the contents of the documents deposited in the office of the public records. If surveys find and make notice of discrepancies in boundaries, then the

[86]Bothin v. Cal. Title Ins. Co., 153 Ca. 718; cited in Stearns at 18 C.A. 3d. 169.

correction of title *must* be made through the deposit of one or more new documents in the public records' office properly executed by those parties having bona fide title and filed in accordance with the applicable code sections.

As one prominent title attorney liked to remark, "You can't map a man out of his title." In other words, a map is a tool of the trade and must be used properly with all consideration given to the exact need of the project. Its loose adaptation to the determination of lines of ownership can defeat its purpose, or, its correlation with other matters and selective application can assure good continuity of title.

Many maps exist outside the usual office of public records as defined in the recording statutes which are helpful in their display of information and are sometimes also a channel to other sources of facts. These are found in such offices as the County Surveyor's office; City Engineer's office; County Flood Control; U. S. Corps of Engineers; State Division of Highways; Railroads; Power, Water, Gas and Telephone Companies; Sanitation Districts, etc. These maps are classified as *quasi-public* maps and they are subject to change without any prior notice to the public. One reason for information being changed on the maps is that the field notes of subsequent surveys furnish additional or different information which is then applied to the map. For this reason, it is undesirable to use or make reference to such quasi-public information — be it maps or field notes — because the moment there is a change, the prior reference is probably void and no one outside of that office knows about it. Of course, this does not negate any value they may have in assisting you in the research of information but do not use them for reference in your descriptions; rather, address yourself to those matters found in the public depository which by law retain their identity over the years.

Conveyed or Described

These words were discussed in the second part of Chapter 3 from the standpoint of what they meant when you read them, but let us consider here the way to adapt them in creating a new description. Usually in referring to a boundary deed the description follows the form of, ". . . . to the west line of the land described in deed to W. R. Brown recorded March 12, 1962 in Book 459, Page 72, of Deeds . . . . ." This means you are tieing to that west line *as it is described* in that deed and properly so. This is the preferable way to use such a reference.

If, on the other hand, you refer ". . . . to the westerly line of the land conveyed . . . .," remember that that may represent more-or-less than the land described and therefore your boundary may be different than you have indicated. See Figure 81.

Where a street in a subdivision is boundary to a lot, normally the lot carries with it the underlying interest in the street. This is a simple illustration of where the land described is the lot but the land conveyed is more being the lot plus the underlying interest in the street.

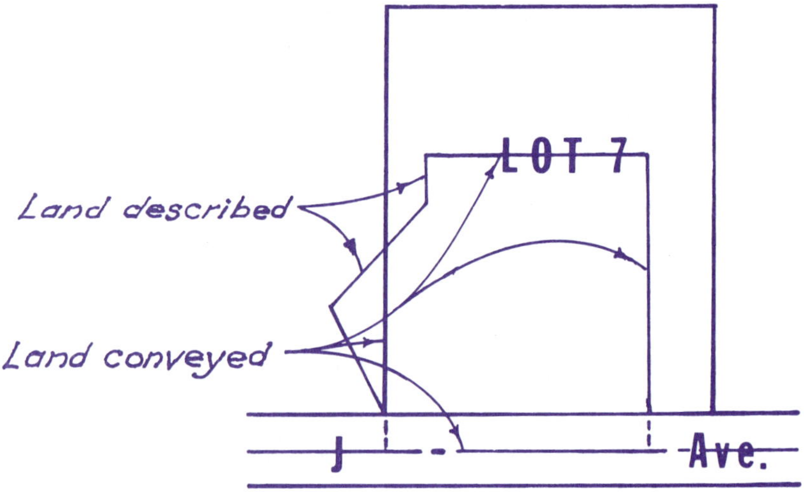

Fig. 81

An entirely different situation exists, however, when you are dealing with a reference to "land conveyed," *which is already a matter of public record.* If you intend to tie to a recorded deed as a boundary which describes your required line as ". . . . the westerly line of the land conveyed to . . . .," you cannot change that connotation. You must consider it as questionable until the position of the true line of conveyance can be established through a careful search of the records behind it. Your research may even result in a stalemate in which case you should seek the advice of an attorney or ask a title company what they would insure and then use their line as your boundary.

Reference to "N.A.P."

Sometimes the letters "N.A.P." are used alone while other times the entire meaning, "Not A Part Of This Subdivision," is printed in the area of a map which is bounded or surrounded by lots of the subdivision because it was the old homesite excepted from the sale. See Figure 82. This isolated area usually has the bearings and distances shown on all of its boundaries the same as if it were one of the lots in the tract. For this reason, it is simpler to describe it by reference to the map rather than by metes and bounds.

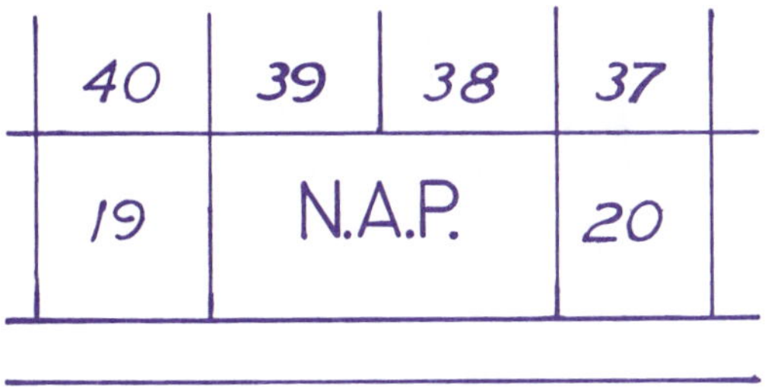

Fig. 82

The principal item to remember in this case is that the title of ownership for the "N.A.P." area is NOT the name or number of the new subject tract, but it is that of the *caption* on the new tract map out of which the new tract came. For example, if you subdivide a portion of Lot 2 in Section 5, T 3 S, ½ 4 E, ____ Meridian, etc., with the old home site left out and shown as "N.A.P." on the new map of the HOYT TRACT, then your description should read thusly:

> That portion of Lot 2 in Section 5 T 3 S, R 4 E ____ Meridian, etc. in the City of ______________________________, County of ______________________, State of ______________________ shown as "N.A.P." on the map of the HOYT TRACT recorded in Book ____, Page ____, of Maps in the office of the County Recorder of said County.

In using this type of description, always adapt your caption to the same information as contained in the CAPTION *on the new tract map.*

Multiple Parcels

Some situations require a number of parcels such as one for the fee area, several for associated easements and perhaps one for separate parking facilities. In these instances, you have no doubt seen them conglomerated in different ways, all under one caption. In many cases, this method is most undesirable because too often one of the parcels is lifted out and put on a separate document *without the caption,* whereby leaving off essential information and this could even void the conveyance for want of sufficiency and certainty.

Another problem also arises when one parcel is lifted out of a group and it refers to a point or a line "in Parcel 2 above," so without that other parcel, the description of the one removed is questionable if not void.

Use of, Instead of Adding to

For the benefit of all concerned, it is advisable and preferred that each and every parcel have its own caption and be able to stand on its own merit; then, if anyone chooses to pick one out of a group to use on a separate document, it will be sufficient unto itself. In other words, use more captions instead of adding more parcels to one caption.

Avoidance of Conflict

The same application of "no conflict" needs to be applied to each parcel in a group as is applied to adjoiners. In writing descriptions for a series of parcels, one of the common occurrences is a lack of conformity created by traveling in opposite directions along the same line and with insufficient ties to hold its position. The fundamental form which will assure the coincidence of lines of adjoining parcels is to describe the common line in the same direction in each discription and use the appropriate corresponding ties in each one. The following are recommended forms of discriptions for Parcels 1, 2 and 3 shown in Figure 83. Each one is written in directions opposite to each other in order to have them follow the *same* direction on lines common. Notice the flat ties used to *hold* certain points.

> **Parcel 1:**
> That portion of Rancho Pauba in the county of Riverside, state of California, as per map filed in book 49, page 59, of Records of Surveys in the office of the County Recorder of said County described as follows:
> Beginning at the most northerly corner of Lot 1 in Tract No. 7821 as per map recorded in Book 36, Page 79, of Maps in the said office of the County Recorder, being also the southerly extremity of the westerly line of the land described in Parcel 1 in Instrument No. 127437 recorded November 8, 1965 in said office of the County Recorder; thence along the said westerly line N 14° 32′ 25″ W 1125.52 feet to Point "A" of the land described in Parcel 3 in Instrument No. 130303 recorded December 30, 1970; thence along the center line of the 60-foot easement for Road and P. U.

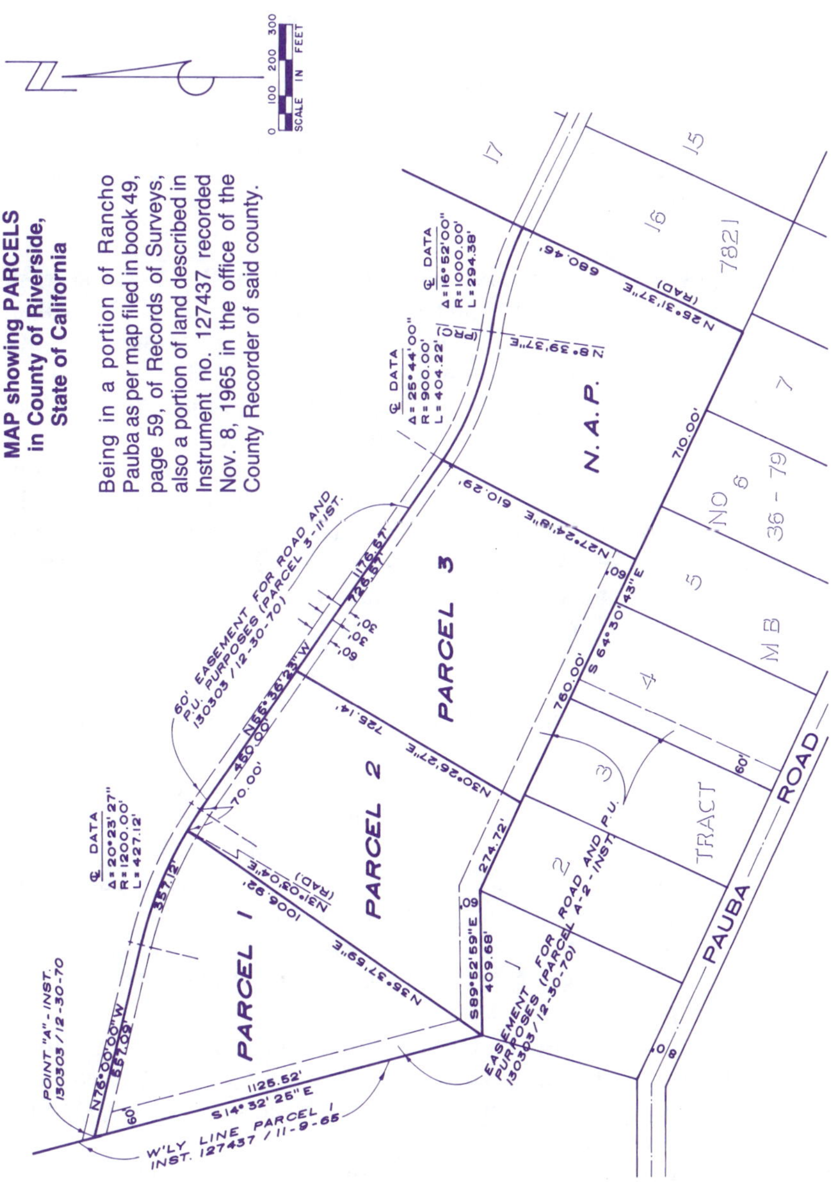
MAP showing PARCELS
in County of Riverside,
State of California
Being in a portion of Rancho Pauba as per map filed in book 49, page 59, of Records of Surveys, also a portion of land described in Instrument no. 127437 recorded Nov. 8, 1965 in the office of the County Recorder of said county.
0 100 200 300
SCALE IN FEET
PARCEL 1
PARCEL 2
PARCEL 3
N. A. P.
60' EASEMENT FOR ROAD AND P.U. PURPOSES (PARCEL 3-II INST. 130303/12-30-70)
EASEMENT FOR ROAD AND P.U. PURPOSES (PARCEL A-2-INST 130303/12-30-70)
W'LY LINE PARCEL I INST. 127437/11-9-65
POINT "A" - INST. 130303/12-30-70
℄ DATA Δ=20°23'27" R=1200.00' L=427.12'
℄ DATA Δ=25°44'00" R=900.00' L=404.22'
℄ DATA Δ=16°52'00" R=1000.00' L=294.38'
N76°00'00"W 557.09'
S14°32'25"E 1125.52'
N35°37'59"E 1006.92'
N31°03'04"E (RAD.)
N30°26'27"E 725.14'
N27°24'18"E 610.29'
N25°31'37"E (RAD) 680.46'
N28°39'37"E
N55°36'23"W 450.00'
S89°52'59"E 409.68'
S64°30'43"E
357.12'
274.72'
760.00'
710.00'
70.00'
176.57'
725.57'
PAUBA ROAD
TRACT NO 7821
MB 36-79
1 2 3 4 5 6 7 15 16 17

Fig. 83

Purposes described therein S 76° 00′ 00″ E 557.09 feet to the beginning of a curve concave to the southwest and having a radius of 1200 feet; thence southeasterly 357.12 feet along said curve to a point northwesterly 70.00 feet along said curve from the northwesterly terminus of that course in said Parcel 3 cited as "N 55° 36′ 23″ W 1176.57 feet"; thence S 35° 37′ 59″ W 1006.92 feet to the point of beginning.

Parcel 2:

That portion of Rancho Pauba in the county of Riverside, state of California, as per map filed in book 49, page 59, of Records of Surveys in the office of the County Recorder of said County described as follows: Beginning at the most northerly corner of Lot 1 in Tract No. 7821 as per map recorded in Book 36, Page 79, of Maps in the said office of the County Recorder, being also the southerly extremity of the westerly line of the land described in Parcel 1 in Instrument No. 127437 recorded November 8, 1965 in said office of the County Recorder; thence along the north line of said Lot 1, S 89° 52′ 59″ E 409.68 feet to the most easterly corner of said Lot; thence along the northeasterly line of said Tract S 64° 30′ 43″ E 274.72 feet to a point northwesterly 760.00 feet from the most westerly corner of the area labeled "N.A.P." on the map of said Tract; thence N 30° 26′ 27″ E 725.14 feet to a point on the center line of the 60-foot Easement for Road and P. U. Purposes described in Parcel 3 in Instrument No. 130303 recorded December 30, 1970 in said office of the County Recorder which point is southeasterly 450.00 feet from the northwesterly terminus of that course in said Parcel 3 cited as "N 55° 36′ 23″ W 1176.57 feet"; thence N 55° 36′ 23″ W 450.00 feet to said terminus and the beginning of a curve concave to the southwest having a radius of 1200.00 feet; thence northwesterly 70.00 feet along said curve; thence S 35° 37′ 59″ W 1006.92 feet to the point of beginning.

Parcel 3:

That portion of Rancho Pauba in the county of Riverside, state of California, as per map filed in book 49, page 59, of Records of Surveys in the office of the County Recorder of said County described as follows:

Beginning at a point on the northeasterly line of Lot 2 in Tract No. 7821 as per map recorded in Book 36, Page 79, of Maps in said office of the County Recorder, which is distant along said northeasterly line and its southeasterly prolongation northwesterly 760.00 feet from the most westerly corner of the area labeled "N.A.P." on the map of said Tract; thence N 30° 26′ 27″ E 725.14 feet to a point on the center line of the 60-foot Easement for Road and P.U. Purposes described in Parcel 3 in Instrument No. 130303 recorded December 30, 1970 in said office of the County Recorder, which point is southeasterly 450.00 feet from the northwesterly terminus of that course in said Parcel 3 cited as "N 55° 36′ 23″ W 1176.57 feet;" thence S 55° 36′ 23″ E 726.57 feet to the most northerly corner of the said area labeled "N.A.P." on said Tract map; thence S 27° 24′ 18″ W 610.29 feet to the said most westerly corner; thence N 64° 30′ 43″ W 760.00 feet to the point of beginning.

This is not the only way to write these parcels because each could be written in the opposite direction *provided* that the comparable ties common to each pair are held and the direction of travel is the same along the line common to each pair.

Of course, with each description, when placed on a document for the sale of that parcel, there should be another parcel granting the appurtenant easements for Road and P.U. Purposes. Since those easements are on record, they can be conveyed by reference to the recorded document, or, described as they are in the document.

A different situation arises when a large area has been cut into several parcels and it is necessary to determine if they are properly correlated with each other and with the exterior boundary. Figure 83a shows the original large parcel A. Figure 83b shows how the

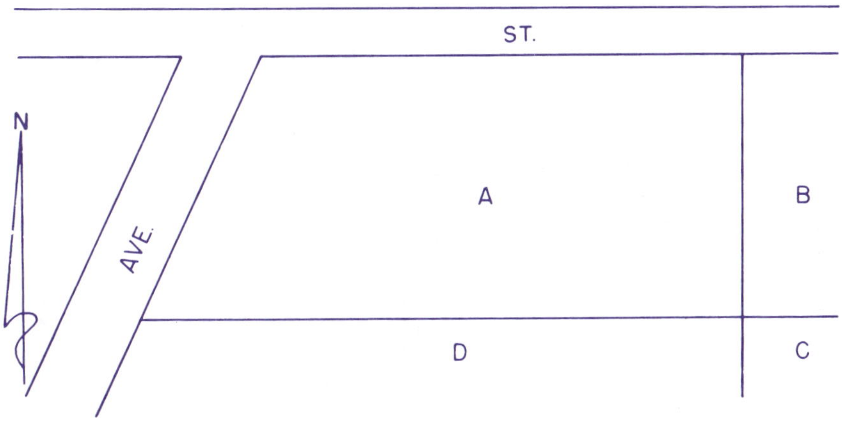

Fig. 83a

owner intended to have the 6 parcels fit completely within the boundaries of parcel A but from reading the descriptions on the documents on record for the 6 parcels, it was necessary to change the owner's

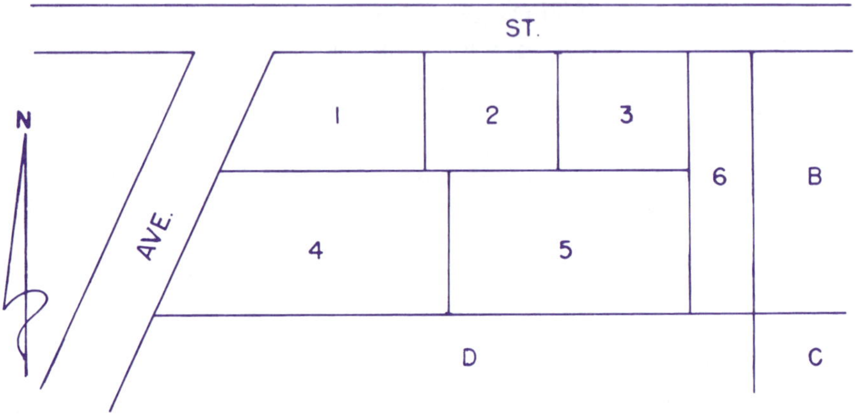

Fig. 83b

map to the configuration shown in Figure 83c. The descriptions for both sub-parcels 1 and 2 overlapped each other; 4 and 5 overlapped on the southerly portion of their common line; although 5 and 6 agreed with each other, they left a gap between them and 2 and 3; finally, 4, 5 and 6 encroached onto the adjoining parcels B, C and D. All of this was the result of descriptions that were ambiguous and insufficient.

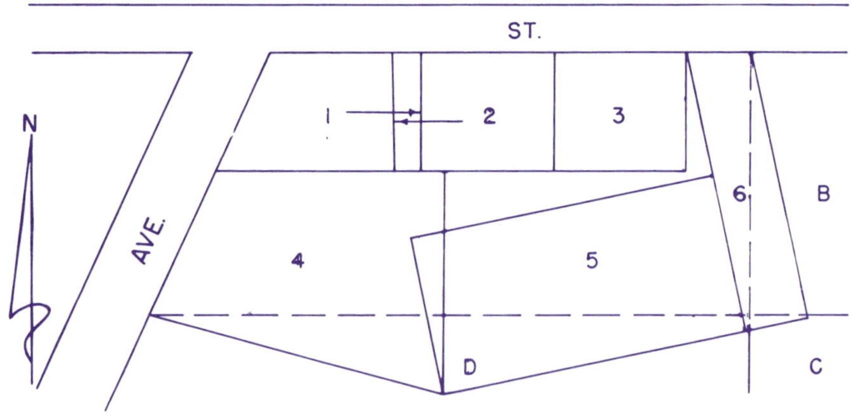

Fig. 83c

The numbering of all areas as shown in Figure 83d helps to segregate the assignment of the different title interests for the purpose of delineating proper legal descriptions to bring the record into alignment with the original plan of the owner of parcel A.

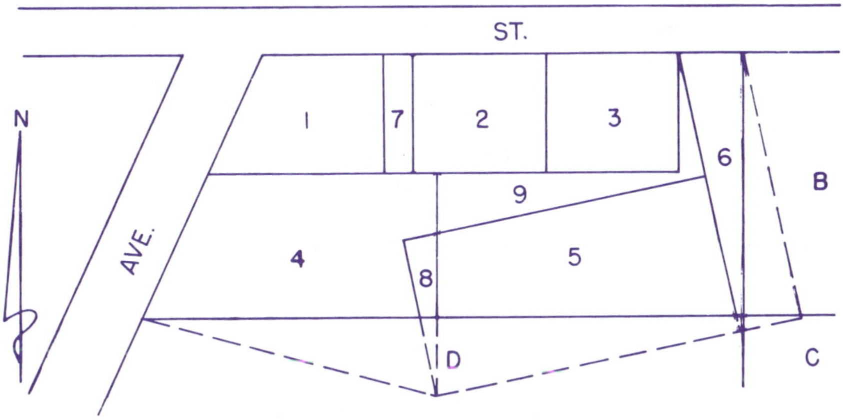

Fig. 83d

One of the problems in this type of situation is with those interests other than fee title, i.e., mortgages, leases etc. In the case of sub-parcels 7 and 8, all encumbrances will need to be re-executed in conformity with the final disposition of the fee.

Since sub-parcel 9 remains in the owner of parcel A, there will be required additional grants from him. The owners of sub-parcels 5 and 9 will need to grant to the owner of sub-parcel 6 areas which together will conform with the original plan for 6.

Although no title could pass to any area described outside of parcel A, it would be a courtesy to the owners of parcels B, C and D to also record a quitclaim on those encroachments.

The lesson to learn from this is twofold: one is that when you have one parcel amongst several, do not assume that they all fit, be sure to draft each and every one of them exactly as the words in each document describe them; the other is to check all encumbrances too and make certain that they conform with any re-elignment.

Lines of Title vs. Monuments

Some discussion has been had regarding monuments and their acceptance or non-acceptance. In writing descriptions, we again have the distinction to make between using, or perhaps revising, existing descriptions and creating all new descriptions. Insofar as monuments vs. lines of title are concerned, three principal items need to be remembered and considered. 1 — Are the monuments specifically cited in the description? 2 — Were the monuments set some time before the description was written?; Were they set at the time of writing the description?; Or were they set an appreciable time after the creation of that description? and 3 — What were the intentions of the parties?

By the rules of construction, a cited monument carries high priority so if you are creating a description and certain landmarks have been agreed upon between the concerned parties, incorporate those monuments into your description by tieing to each one as you proceed around the traverse. If, on the other hand, you are adapting or revising an existing description which does *not* refer to monuments and you want to add references to them, DO NOT add them unless you have absolute survey proof, *together with* enough facts derived from your research of record title and off-record matters to show that they are coincident.

To assume that a hitherto uncalled-for monument, whose position does not agree with the line described in the deed, is acceptable for the control of that line is to assume too much. Such acceptability is not always upheld.[87] (New monuments have been held acceptable and binding, however, in one case where purchasers relied upon iron stakes set for lot corners in a tract.[88])

The reason why monuments which have been set a period of time prior to the writing of the description may be of questionable status is that there could have been a change in the lines agreed upon for the final sale of the land without correlative changes of the monumentation.

It is generally considered that when the description is created at the same time the monuments are set, there will be no conflict between them, or if it develops that a conflict exists, the monuments called for will hold on the basis of representing the surveyor's footsteps and the understanding of the parties involved. The exception is if it can be proved that any monument was set erroneously.

Where monuments are set some time after the description is writ-

[87]McKinney v. Doane, 155 Mo. 287, 56 SW 304; Powers v. Jackson, 50 Ca. 429; Crandall v. Mary, 67 Ore. 18, 135 Pac. 188

[88]Arnold v. Hanson, 91 CA 2d. 15, 204 P 2d 97.

ten and with no reference to them, they will hold if they conform but not in the case of non-coincidence as per footnote 87 supra.

If extrinsic information constituting admissible evidence is introduced which shows a situation contrary to the application of the rule, then it (the rule) is subject to such adjustments as may be necessary to comply with the proven intent of the parties.[89]

Therefore, in the final analysis of writing descriptions, use references to monuments where applicable but do not try to incorporate them where they cannot stand the test of conditions referred to above. Remember that lines of title usually prevail on the basis of the public record because of its public notice and for all of the information contained therein.

Lines of Title vs. Occupation

Here we are distinguishing occupation from monuments by talking about things that are built across lines of title such as buildings, fences, walls, etc. As discussed under boundaries, the matter of acquiescence, agreement lines and prescriptive claims, each have their respective modus operandi for making it possible to move a line of title from an old position to a new position in a way that will perfect the title of the land between the former line and the new line, thereby making it marketable.

When occupation across a line of title exists for a time longer than the time specified in the statute of limitations applicable to the occupation, various laws assist the claimant in establishing certain rights. However, when an adverse occupation is relatively new and the facts are revealed by survey, then the adjoining owners need to reach an agreement, or sue, as to the solution of the problem.

In either situation, the scrivener is called upon to take the facts presented, usually from a field survey, and create a proper description to cover the area in question. The primary objective to remember is: clearly set forth the lines of title and distinguish from them the lines of occupation.

Another aspect of this subject is the matter of lines of title compared with lines of occupation after a major earth disturbance. To accommodate the many descrepancies in the San Francisco area, the McEnerney Act was passed by the State Legislature which provided for quiet title action on and by individual holdings. Claims were based on actual survey of the areas showing lines of occupation which

[89]Miller v. Grunsky, 141 C. 441, 66 P. 858.

resulted from the earthquake. The decrees issued from these actions established the new lines of title to prevail from there on. The constitutionality of this act was upheld by the California Supreme Court in Title Document Restoration Co., v. Kerrigan, 150 Ca. 289, 88 P. 356, and further supported by the United State Supreme Court in American Land Co. v. Zeiss 219 US 47.

In Alaska, legislation was passed providing for court action "Relating to establishment of land boundaries affected by earthslides." This correction of existing public records by a "substitute plat" by "judicial approval" to be accomplished in a single action in rem although a permissable plaintiff entity may bring a separate action.

On this basis, also dependent on a field survey of the facts of occupancy, the separate lines of title in groups of one or more blocks of ownerships are established by decree. A new map covering the area affected by the action approved by the judgement must be properly filed as a substitute for maps previously filed.

The more recent legislation passed in California, known as the "Earth Movements Disasters Act" calls for the operation of the single action in rem but also permits of a separate action under certain conditions by court approval. Its coverage is much broader, however, in that it concerns itself with, "boundaries .... disturbed by earth movements such as but not limited to slides, subsidence, lateral or vertical displacements or similar disasters caused by man or by earthquakes or other acts of God."

Action may be initiated by a city, a county, an entity or person owning or having an interest in or lien upon the land affected by the disaster if granted by the court. As a result of the court's action and by its decree, an official map must be filed as a substitute for all previously filed maps covering the area affected by the action. The certified copy of the judgement shall be recorded and shall constitute constructive notice of the findings therein and of the official plat or plats referred to therein.

As to descriptions of land affected by earth movement, the record information will stand until proceedings of suits are concluded which change those record lines to the appropriate new position of occupation. This alteration may be accomplished through court action in the case of large groups of parcels, or in some instances, agreements between individuals properly executed and recorded may suffice.

It is interesting how time seems to cure some decrepancies. In the case of the 1933 Long Beach earthquake, the alignment of a half-mile of straight road in the adjoining city of Wilmington had just been staked. After the quake that line of stakes followed a parabolic curve with a maximum offset from the straight line of about six feet. Some twenty-five years later, that curved line had returned almost entirely to a straight line.

Positive Calls vs. Afterthoughts

Too many times a statement is made in a description intended to qualify a point, a line or a condition but that statement follows sometimes several lines after the thing it qualifies. Such a placing of that type of declaration gives the impression that it was just an afterthought, a supplementary unimportant matter that just happened to come to the mind of the scrivener after he passed the point where it was pertinent. In this way the qualification loses its punch and might even be declared ineffective.

If a point or a line is to be controlled by a tie to a physical monument, a line on a map, a boundary deed, etc., then make that call a positive statement and place it in the description at the position to which it applies, not at a later point.

Condominiums

Evidence of this type of ownership has been found in history as far back as the sixth century, B.C., in Rome.

Black's Law Dictionary defines it as "System of separate ownership of individual units in a multiple-unit building. Susskind v. 1136 Tenants Corp., 251 N.Y.S. 2d. 321, 43 Misc. 2d. 588." Although a condominium owner is sometimes referred to as a "joint owner," he is in fact a single owner of the fee of a cube in space as are his neighbors in the same complex.

Even concerning those elements of common use, the walls, foundation, roof, hallway and elevators in some cases, swimming pools, greenbelt areas, driveways, etc., the individual fee owners of cubes in space are not necessarily "joint owners" of the common elements but are in fact owners of an individual undivided interest in them or in some cases a fractional interest.

The limits of the common description applied to the volume of air in space for a condominium are the surface of the walls, the lower surface of the ceiling and the upper surface of the floor. To put this into a three-dimensional metes and bounds description for each and every unit, you must agree, would be very cumbersome and difficult. Again, using the device of a map and diagrammatic plans is the simpler and safer method of description, but before going into these details, let us consider and list the items to which we must address ourselves in the overall picture.

Besides the fee title to the unit itself, there may be either an undivided interest in and to the common areas, based upon the total number of units, or a fractional interest based upon a proportionate division of the square footage of floor area or the price of the unit in proportion to the total price. There may also be a grant of an interest or fee title in and to some additional elements such as a garage, patio, storage space or balcony.

There must be a primary instrument of record covering the common area(s) by which reference to these items can be made. It can be a metes and bounds description, an already recorded lot or parcel on a map or a new subdivision map of one or more lots. In some cases, the units are shown within the common area, while, in other cases, they are set forth on separate sheets called diagrammatic plans. The units, along with any correlative elements, are all assigned their respective identifiers. Wall space between units is shown and dimensioned and usually at least one of the exterior walls is oriented and tied to the boundary of the common area. These plans may be filed on full map size sheets with the tract map or reduced to legal size and filed in the Official Records, either with or separate from the Declaration of Restrictions.

Information concerning the vertical limits can be disseminated in one of three ways: 1 — Elevations are shown on the plans for top of floor and bottom of ceiling of each unit or element; 2 — A typical cross section is illustrated with the elevations shown at floor and ceiling; 3 — A tabulation of all the elevations is put on a sheet of the plans. The control of this vertical information is based on bench marks established, logged and maintained by responsible offices.

In order that the plans will reflect conformity of facts with the physical structures, it is recommended that they be recorded after construction of the units and that the engineer's or surveyor's certificate confirm the agreement of those plans with the structure. If the map is recorded prior to the erection of improvements, then the surveyor or engineer should furnish a certificate properly acknowledged that the improvements have been completed according to the map, if it is so, or, if not, set forth the differences in the certificate. If the variations are major, then a corrected map should be filed.

All of these processes are for the benefit of having a reliable description in the document of conveyance.

The following are examples of descriptions to use with the various conditions mentioned above.[90]

EXAMPLE 1:

(When using a conventional surface subdivision map and the units are shown within one or more of the lots in the tract. Sometimes everything is included within a one-lot subdivision, while at the other extreme, many lots are designed into the tract carrying perhaps four or six units each with several lots set aside for recreational and utility purposes. In the latter case, more than one lot will be mentioned in Parcel 1.)

"A condominium comprised of:

PARCEL 1: An undivided ___ interest in and to Lot ___ of Tract No. ___, in the City of ______________, County of ______________, State of ______________, as per map recorded in Book ___ of Maps, at Page ___, in the office of the County Recorder of said County.

PARCEL 2: Unit ___ as shown upon the map referred to in PARCEL 1 above."

EXAMPLE 2:

(When the diagrammatic plans of the units are on file in the Official Records and not attached to the tract map.)

"A condominium comprised of:

PARCEL 1: An undivided ___ interest in and to Lot ___ of Tract No. ___, in the City of ______________, County of ______________, State of ______________, as per map recorded in Book ___ of Maps, at Page ___, in the office of the County Recorder of said County.
EXCEPTING THEREFROM Units ___ through ___ as shown upon the condominium plan recorded in Book ___ at Page ___ of Official Records of said County.

[90]Recommended forms by California Land Title Association.

PARCEL 2: Unit ___ as shown upon the condominium plan referred to in PARCEL 1 above."

EXAMPLE 3:

(In those cases where more than one condominium element is to be conveyed to the owner in severalty, such as a residential unit, a balcony, and a storage area.)

"A condominium comprised of:

PARCEL 1: An undivided ___ interest in and to Lot ___ of Tract No. ___, in the City of ______________________, County of ______________________, State of ______________________, as per map recorded in Book ___ of Maps, at Page ___, in the office of the County Recorder of said County.

EXCEPTING THEREFROM Units ___, ___ and ___ as shown upon the condominium plan recorded in Book ___ at Page ___ of Official Records of said County.

PARCEL 2: The unit consisting of Elements ___, ___ and ___ as shown upon the condominium plan referred to in PARCEL 1 above."

EXAMPLE 4:

(Where, in addition to a unit, an exclusive right of possession of elements of the common area, such as patios, balconies or garages are conveyed.)

"A condominium comprised of:

PARCEL 1: An undivided ___ interest in and to Lot ___ of Tract No. ___, in the City of ______________________, County of ______________________, State of ______________________, as per map recorded in Book ___ of Maps, at Page ___, in the office of the County Recorder of said County.

EXCEPTING THEREFROM the following:

(a) Units ___ through ___ as shown upon the condominium plan recorded in Book ___ at Page ___ of Official Records of said County.

(b) The exclusive right to possession of all those areas designated as balconies, patios, storage areas, and garages as shown upon the condominium plan above referred to.

PARCEL 2: Unit ___ as shown upon the condominium plan above referred to.

Where the condominium project occupies a parcel of land described by metes and bounds, Examples 2, 3 and 4 are adaptable by using the following form for the first part of Parcel 1:

An undivided ___ interest in and to the land described as follows:
(use metes and bounds description here)

Follow with the EXCEPTIONS applicable to your condition and add Parcel 2 as before.

Sometimes the Declaration of Restrictions is the medium whereby a parking and/or storage space is provided by assignment to the occupant of Parcel 2.

Along with the conveyance of these Parcels on the Grant Deed, reference to the Declaration of Covenants, Conditions and Restrictions is also given with its recording information.

These forms and the application of principles can be used for not only residential units but also industrial, commercial, medical and other classifications of condominiums.

Coordinates

There were no coordinate controls or grids in the early days to which to tie survey points. The earliest survey work on base lines for the beginning of triangulation — which led to the use of coordinates — was in 1816 and this was interrupted for 14 years by a lack of funds.

Land titles stand on the framework of legal descriptions developed from, and their continuity supported by, surveys. In some cases, the courts have held that the marks on the ground constitute the survey while the courses and distances given in a conveyance are only evidence of a survey.[91]

Surveying relies on the science of mathematics in the form of measurements of distances and angles for input information. But the determination of survey lines and points representing ownership is controlled by legal doctrines concerning boundaries, monuments, topography, priority in time and facets thereof. This is why the professional surveyor must be well versed in both the discipline of measurements *and* the knowledge of legal principles involved in the perfection of title to land.

[91]Millerman v. Megaritty, 242 SW 757, (Tex.); Andrews v. Wheeler, 103 Pac. 144 (Cal.).

11.40

Coordinates are an excellent tool for many purposes and they assist the surveyor in numerous ways, both in the field and office. It is possible to devolve coordinates from a triangulation network down to any point desired and because of the simplicity of expanding coordinates from point to point, it has been suggested that every corner of every parcel of land be assigned a pair of coordinates, but let us break away the mechanics of such an operation from the legal status of corners of property.

What legal value does a point with coordinates possess if it is not the correct corner of ownership? There are many corners and lines of title under dispute awaiting court decision, an agreement or some sort of legal action. Are they to be assigned coordinates which must be subsequently changed?

Some states have provided for the acceptability of coordinates assigned to parcel corners, others have merely legislated the creation and development of a state plane coordinate system, while a few have been specific in recognizing coordinates and establishing a statewide system but denying any legal status of coordinates over and above ties to matters of record insofar as they may control the boundaries of ownership.

This does not mean you cannot write descriptions with coordinates but from the standpoint of title, any such application should be on a parenthetical basis so as to retain the certainty of legal control. Again, consult *your* state codes to determine the extent to which coordinates have been established, the form of designation assigned to them and their legal connotation.

The assignment of coordinates to a point or monument by a surveyor, a photogrammetrist or a draftsman merely identifies an accurate method for reestablishing that point on the ground or on paper; it does not make a determination of its title value. Until or unless the coordinates of such a point are tied to, or made a part of, matters of record in a description, their value is no greater than any other dimension.

It has also been suggested that all monuments be disregarded and that co-ordinates be the sole label and responsibility of property corner designations. Many decisions by the courts in land boundary cases are based on surveys and those surveys in turn are based on monuments and metes and bounds descriptions. So, regardless of anyone's attitude toward monuments, it is necessary to live with them where necessary as embraced by the law until it is changed.

This is why surveyors, or anyone else, cannot ignore *properly* established monuments. Even if the law were changed to disregard all monuments, it is extremely doubtful that it would be retroactive because it would cause too many economic problems. Therefore, all those previously acknowledged would still have to be honored. Furthermore, many people would become very skeptical of coordinates *they* could not understand in lieu of the physical monuments they could see.

Another objection to the sole use of coordinates is that one or more errors in the figures can throw the entire area out of title and you have no warning of it. In other words, a miscopy or transposition of a number could be made accidentally and there would be no signal to inform anyone of the mistake until extrinsic conditions brought it to light.

With legal descriptions, you have built-in checks with ties and monuments to which lines must conform. Properly written descriptions are inherently correct insofar as legal title is concerned. The legally allowable variation in measurement by ties permits the survey and ground conditions to be brought into agreements with each other and with title.

In writing a description incorporating coordinates, the fundamental specifications are that they refer to the correct zone, if there is more than one, the state designated nomenclature of the Coordinate System and if it is Grid or some other type of reference.

Although a single coordinated point in a mathematically closed description would position the property (provided the bearings were on the same basis as the coordinates used), it is preferable to include at least two coordinated points. It is not necessary to assign coordinates to every point inasmuch as they can be derived from the traverse; rather, it would be advisable to choose pertinent control points that have been tied by actual survey to, or can be accurately computed from, higher order controls. The following description illustrates the application of coordinate inclusion.

EXAMPLE:

That portion of the SE ¼ of the SE ¼ of Section 15, T ___ S, R ________ E, of the ______________Meridian in the County of ____________________________, State of ____________________________, according to the Official Plat of said land filed in the District Land Office described as follows:

Beginning at a nail set in a square cut stone on the south line of said Section in the middle of Three-mile Road being the southwest corner of the land described in deed to Su San Kim recorded in Book ___, Page ___, of Deeds of said County at a point N 89° 48′ 16″ W 400.00 feet from the southeast corner of said Section, being a marked stone having established grid coordinates of (X)______, (Y)______, of Zone 2 of ___State Coordinate System; thence along the westerly line of said Kim land (bearings based on grid meridian of said Zone 2) N 28° 17′ 32″ " E 200.00 feet; thence N 66° 32′ 51″ W 331.64 feet to a point on the east line of the WW Tract as per map recorded in Book ___, Page ___, of Maps in the office of the County Recorder of said County distant along said line S 0° 10′ 24″ E (record bearing of said east Tract line shown as SOUTH) 102.68 feet from a 2" (O.D.) iron pipe at an angle point in said Tract line having grid coordinates of (X)______, (Y)______, of said Zone 2; thence along said Tract line S 0° 10′ 24″ E to said south line of Section 15; thence S 89° 48′ 16″ E to the point of beginning.

Clarity

Through all of your writing, strive to express yourself in ways that will leave a clear picture in the mind of the reader. After having written a description, set it aside momentarily, take your attention off of it while you do something else and then go back and read it *as if it had just come to you from someone else.* See if it leaves the correct impression on you. Are the parts in proper form? Do the Exceptions follow in good order? Is each statement, each reference sufficient in itself? If the parts are sufficient, the whole will be sufficient. Are the beginnings well explained? Is the base of bearings properly established? Are the ties and reference material suitable? Is there any conflict?

Another safety valve for checking is to read your description again after having dictated it to your secretary and after she types it. She might have received a different impression and typed something foreign to your intention! It pays to review it.

Metrication

There is no problem concerning the use of metric units for distances in lieu of english units in descriptions. There will be no change in bearings and angles because they will continue to use the sexgesimal system of degrees, minutes and seconds.

Problems

1. Name four general formats used in creating a legal description.
2. When are rules of construction needed?
3. Can you create a valid description with no measurements of any sort? Explain.
4. Is "fractional" lot a correct statement of fact? Explain.
5. Can you safely use "EXCEPTIONS" in the caption of a description? Explain.
6. When is a description sufficient?
7. Name six conditions from which a base of bearing can be derived.
8. How many base of bearings can you have in one description?
9. Are you to assume that every tie made in a description is valid and unquestionable?
10. Name three kinds of "Notice" as applied to references in legal descriptions.
11. When a map does not come under the classification of "constructive notice" by statutory rules, how do you accomplish the incorporation of desired information on the map into a constructive notice channel?
12. Can you safely use quasi public records for ties and references in a description? Explain.
13. List at least six sources of maps which would be classified as "quasi public."
14. Does the "land conveyed" normally cover the same area as the "land described"?
15. What is the rule of direction to follow in describing multiple adjoining parcels?
16. Draw a map from the following description. Although it lacks certain information which would be supplied by a survey, it is nonetheless sufficient for the conveyance of title and can be drafted.

    That portion of Lot 10 of FARM LOTS in the county of Riverside, state of California, as per map recorded in Map book 6, page 8, in the office of the County Recorder of said county, described as follows:

    Beginning at a point on the east line of NEW STREET, 60 feet wide, as described in deed to said county recorded in book 752, page 13, of Official Records, North 6° West 110 feet from its intersection with the north line of Tract No.

1082 as per map recorded in Map book 92, page 6, in the said office of the County Recorder; thence North 88° East 80 feet; thence northeasterly to a point on the westerly line of the land described in deed to A recorded in book 685, page 20, of Deeds distant along said line S 10° 30′ E 335 feet from the northwest corner of said land of A; thence N 10° 30′ W 335 feet; thence on the prolongation of the northerly line thereof, being also the southerly line of a Private Street, 30 feet wide, as described in book 452, page 18 of Deeds, North 76° West 120 feet to an angle point therein; thence South 70° West 90 feet to the said east line of NEW STREET; thence South 6° East 520 feet to the point of beginning.

17. Draw a map of Lot 132 of Tract No. 9771. It is 650.00 feet east-west by 750.00 feet north-south with El Myra Ave., 100.00 feet wide on the west side, bearing NORTH. (This will fit an 8-½″ x 11″ sheet using a scale of 1″ = 100′.)
Inside the Lot, draw the parcels described below. If perchance you find more than one position for any line, show both and label each one. Label all parts by reference to the following:

Parcel 1:

That portion of Lot 132 in Tract No. 9771 in the City of Irvine, County of Orange, State of California, as per map recorded in Book 284, page 36 of Miscellaneous Maps in the office of the County Recorder of said County described as follows:

Beginning at a point 300.00 feet East and 200.00 feet North from the southwest corner of said Lot; thence North 235.45 feet; thence N 47° 28′ 34″ E 229.81 feet to a point on the south line of the land described in deed to J. Vandevort recorded January 14, 1972 in Book 1236, page 29 of Official Records,[1] distant 198.22 feet w'ly of the southeast corner of said land; thence along said south line East 198.22 feet to the east line of said lot; thence South 647.5 feet to a one inch iron pipe tagged LS 2856 on the east line of said Lot at the north face of a concrete block wall; thence parallel with the south line of said Lot West 255 feet more or less to the easterly line of the land described in deed to R. Gould recorded June 16, 1971 in

[1]O.R. 1236/29 = North 100.00 feet of Lot 132 in Tract No. 9771.

Book 1182, page 75 of Official Records;[2] thence N 30° 20′ 00″ W 230 feet along the easterly line of said R. Gould to the point of beginning.

Parcel 2:

An easement for Parking, Ingress and Egress over the north 235.45 feet of the south 435.45 feet of the west 300.00 feet of Lot 132 in Tract No. 9771 in the City of Irvine, County of Orange, State of California, as per map recorded in Book 284, page 36 of Miscellaneous Maps in the office of the County Recorder of said County.

RESERVING THEREFROM an easement for Road and Public Utility Purposes to be used in common with others over the north 30.00 feet of the south 435.45 feet of said Lot.

18. Write descriptions for the following parcels *WITHOUT* using metes and bounds for any of them. Create your own caption.

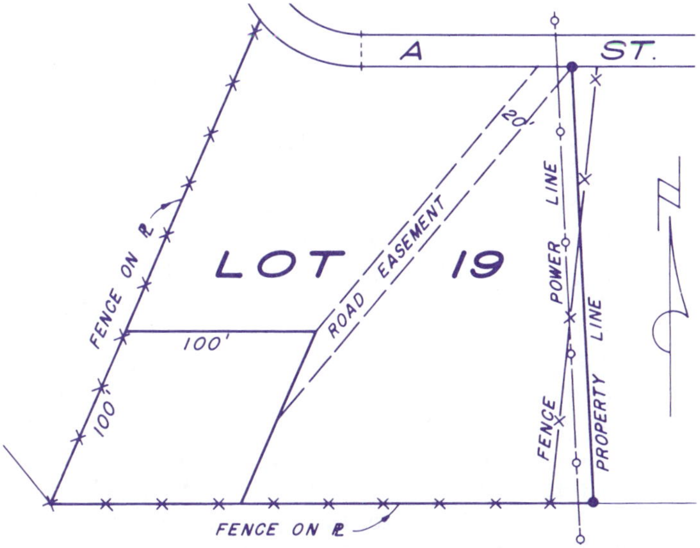

Parcel 1: Describe 100′ x 100′ parcel in SW corner of Lot 19.

Parcel 2: Describe 20′ Road Easement between Parcel 1 and "A" St.

[2]O.R. 1182/75 d.a.f.: Beg. at southwest corner of Lot 132 in Tract No. 9771; thence North 200.00 feet along west of said Lot; thence East 300.00 feet; thence S 30° 20′ 00″ E 230 feet more or less to a point on the south line of said Lot East 395.00 feet from the southwest corner thereof; thence West to point of beginning.

Parcel 3: Describe remainder portion of Lot 19 which lies SE of the NW line of the Road Easement and reserve the Road Easement from it.

Note: Although Lot 19 is shown on a recorded Tract map, *none* of the inside parcels are on record so you cannot tie one to another by reference to the parcel.

19. The following description was excerpted from a recorded document. The interpretation can be questioned; however, it could be written in a better form. Rewrite this description using only one caption and a shorter body without unnecessary repetition.

A portion of the Southeast quarter of Section 17, Township 5 South, Range 10 West, in the Rancho Las Bolsas as shown on a map recorded in Book 51, page 12 of Miscellaneous Maps, in the office of the County Recorder of said county, described as follows:

A portion of the North one half of the North one half of the Northeast quarter of the Southeast quarter of the Southeast quarter of said Section 17, Township 5 South, Range 10 West, excepting therefrom the North 2 acres and the South 1 acre thereof, described as follows:

Beginning at the Southeast corner of tract 4957, as shown on a map recorded in Book 177, pages 23 and 24, Miscellaneous Maps, records of said Orange County, said corner being on the center line of Euclid Street; thence South 88° 09′ 20″ West along the South line of said Tract 4957, being also the South line of the North 2 acres of the East one half of the Southeast quarter of the Southeast quarter of said Section 17, Township 5 South, Range 10 West, a distance of 368.36 feet; thence South 1° 50′ 40″ East a distance of 65.85 feet; thence North 88° 09′ 03″ East a distance of 367.04 feet to the center line of said Euclid Street; thence North 0° 41′ 40″ West along the center line of Euclid Street a distance of 65.86 feet to the point of beginning.

Explain the possible relationship to the north line of the "south 1 acre thereof" and how you would safeguard your client's interest relative to that north line.

Chapter 12
Easements and Rights-of-Ways

While the absolute ownership of land is expressed as a *fee simple absolute*, of which the short form "fee" implies the whole of the meaning, there are various forms of rights, or interests or claims that may be placed upon the land, thereby encumbering the land. These may take the form of rights of access, rights-of-ways and other easements, monetary interests, water rights, mineral interests and various forms of rights for individual benefits such as fishing; more on this later.

It is not the intent here to explain all the ramifications of these encumbrances but to alert you to their existence and use, and discuss the relative forms of descriptions.

Definitions

An EASEMENT, usually for the benefit of one or more individuals, is an interest in the land of someone else and therefore constitutes an encumbrance on another's land. A fee owner cannot own an easement over his own land. An easement carries a limited and non-possessory use by citation. It is created by grant or agreement, expressed or implied. It is transferred subject to the rules of real property. In legal terminology, it is an *incorporeal hereditament.*

An easement, being non-possessory, does not have the protection afforded to a possessory interest such as a tenant under lease. The easement holder has only such control as may enable him to use it as specified in the grant thereof. Ordinarily, this does not exclude others from making use of the same land in a way which does not interfere with the enjoyment of prior rights. In other words, there may be multiple easements over the same strip or area of land but each one must honor those whose rights were acquired previously. An easement granted through the medium of a document is never lost as a result of non-use.

There are both positive and negative easements; the right of ingress and egress would be an affirmative easement; the restriction against erecting buildings on an area reserved for light and air space for the benefit of adjoining land would be a negative easement. See Figure 84.

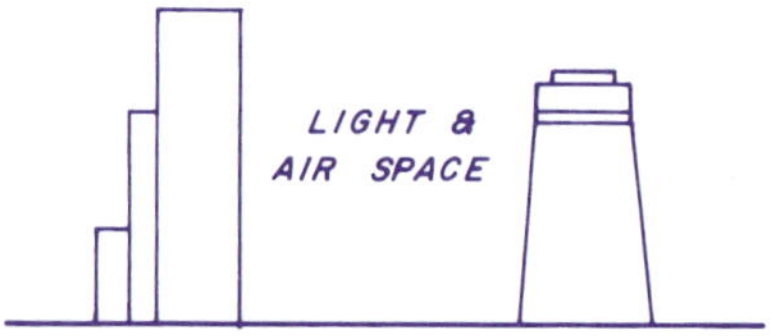

Fig. 84

An easement affects land and therefore the creation of it and the transfer or the interest in the easement must be in writing and express the essentials necessary to ordinary conveyances of land, including a clear description of the easement and grantor's signature. Two exceptions of this manifest, however, when the status of an easement is created by words of covenant between parties concerning a piece of land, or, by the exception of an easement such as for ingress and egress (location not described), but the location of which may be established subsequent to the conveyance.

"An oral grant of an easement creates only a license unless the grantee makes expenditures on the faith of the oral grant, in which case an easement may arise under equitable principles. Wilkes v. Brady, 84 CA 365, 258 P 108."[92]

Land is held as one of two estates in respect to easements: as a *dominant tenement* or as a *servient tenement.* The *dominant tenement* is the land to which an easement is attached or the service is owed. The *servient tenement* is the land which is burdened with a servitude. When one has an easement over adjoining land, it benefits his fee ownership and his land becomes the dominant tenement. The same easement is an encumbrance on the adjoining land which thereby becomes the servient tenement. See Figure 85.

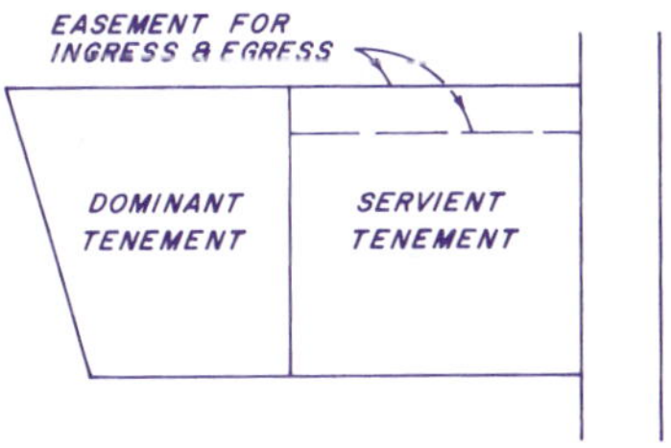

Fig. 85

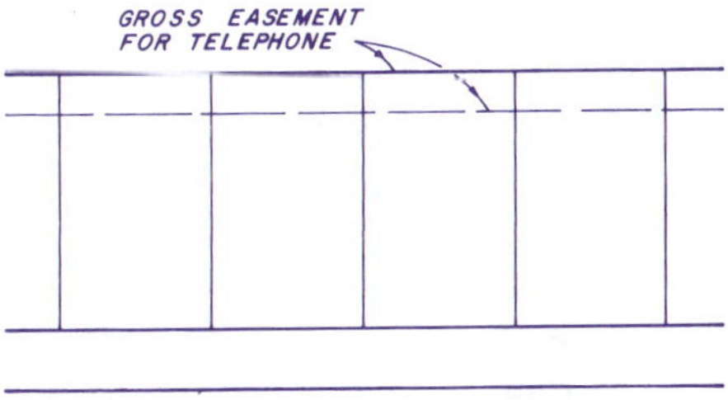

Fig. 86

There are two classifications of easements, *appurtenant* and *gross*. An *appurtenant* easement is created, attached to and for the benefit of the land of the owner of a dominant tenement. A *gross* easement is a right in another's land, which right is not for the benefit of any land owned by the easement holder. The easement for ingress and egress shown in Figure 85 is known as an appurtenant easement for the benefit of and attached to the dominant tenement. An appurtenant easement once established may pass without a specific grant. "An easement for telephone purposes over the North 6 feet of Lots . . . .," would be a gross easement as shown in Figure 86. An easement in gross is not attached to the land, as is an easement appurtenant, but is considered to be a personal or corporate right attached to the easement holder, yet that right does constitute an interest in another's land, which again is the servient tenement.

You, as a scrivener, will not normally be called upon to determine

[92]BOWMAN, Ogden's Revised California Real Property Law, Vol. 2.

the question of whether an easement is *appurtenant* or *gross*. Although you should refer to the codes and the numberous cases concerned with this differentiation for your own edification, any legal determination as to validity or qualification should be made by an attorney or a title company. In the case of appurtenancy, the matter often hangs in doubt as to which particular parcel the easement is appurtenant. Your services may be requested to graphically position an easement in relation to the boundaries of either/or both the dominant tenement and/or the servient tenement. You may also be required to write descriptions of easements, be they en toto or for missing segments.

In regard to easement descriptions, it is important that the *purposes and uses* of the easement be expressed because by rule, it is a right to a limited use or enjoyment of land. For example, if one is acquired for "ingress and egress" and subsequently it is desired to install a water line and power lines, there was no provision made for the latter uses so consequently, the granted easement is insufficient. Furthermore, it is desirable to designate if the easement is appurtenant and if so, to what property.

Although you have read many documents wherein the purpose(s) of the easement has been clearly set forth, such as:

> "A non-exclusive easement for road purposes . . ."
>
> "An easement 10 feet wide for the installation and maintenance of two oil pipe lines . . . . ."
>
> "An exclusive easement for ingress and egress and utilities in and over a strip of land 14 feet in width . . . ."
>
> "A one rod wide easement for a water line and driveway for maintenance . . . . ." etc., etc.,

it is not up to you to create and write such preambles because that is the prerogative and responsibility of a lawyer, or, in some instances, a title company in connection with a specific order.

Overburdening

A word about overburdening an easement. In a simple subdivision of land as illustrated in Figure 87, the use of easements over Parcels V and X by Parcel U is normally acceptable, but if the owner of Parcel U divides it into ten parcels and grants each of them easements over V and X, it would be considered an overburden on the easements over V and X and probably held as unreasonable. The title to the easements held by the owners in subdivided Parcel U is good; it would be a matter of unauthorized increase in burden, opening the door to an injunction against the increased use. There is one case (Sec. 28, C.J.S. 62) where an easement was regarded terminated by the increased burden because, under the circumstances, it was impossible to separate the authorized and unauthorized uses.[93]

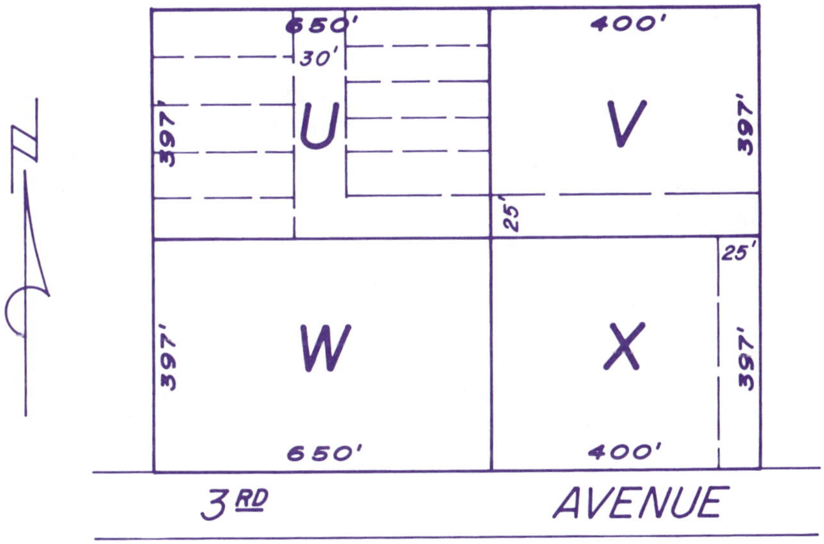

Fig. 87

Rights-of-Ways

Before discussing some forms suitable to easement descriptions, let us consider the relationship and semasiology of *rights-of-way*. Although a right-of-way is technically an easement, it is peculiar in that it is expressly for "passage" purposes; it may be for a railroad, pipe lines, pedestrians, vehicles, a canal, etc. In essence, a grant or creation (possibly without a grant) of a right-of-way establishes a privilege to pass over another's land, under in the case of a tunnel and aerially over in the case of a bridge. It may be for the benefit of an individual, a group or class of persons or the public. Because *right-of-way* is ofttimes used to describe the strip of land itself, questions arise for judicial determination as to whether that strip is in fact an

[93]PRIVATE EASEMENTS from Proceedings of California Land Title Association, May 1951.

easement, or, a fee subject to the right of a specific use. Another definition for right-of-way is a composite of strips of land over which an easement passes. Although a *right-of-way* is technically an easement, when the wording taken as a whole throughout the document is unclear as to intent, it is better for the court to decide.

Strip Easements

As the name implies, a strip of land is created either in toto or in segments, either in the status of an easement or a fee limited to certain rights as stated above to accommodate a certain use.

The strip "in toto" description is easy to create because after making a tie at one end, it takes off across country with little or no concern for intermediate ties to title lines until it ends somewhere. Problems sometimes arise when one or more owners along the way want finite determination of their boundary line in relationship to the easement *prior* to conveyance. At other times problems arise when it is found

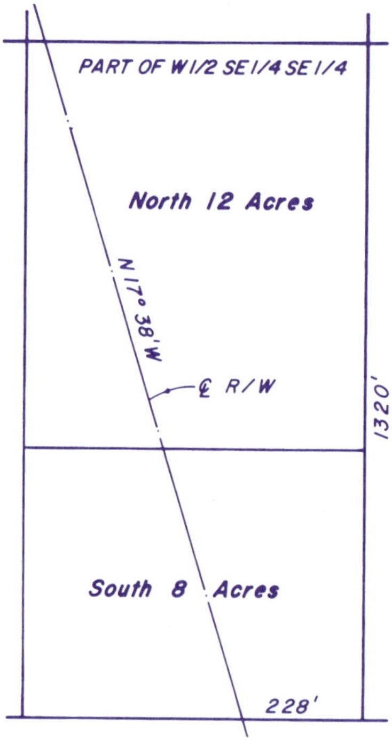

Fig. 88

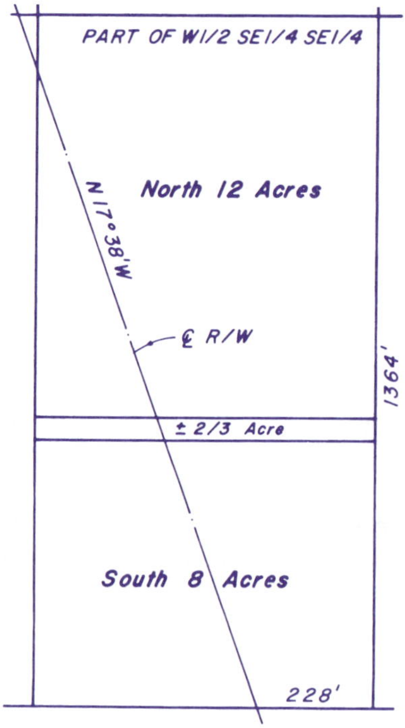

Fig. 89

later by field location that the right of way in fact cuts across corners of other land for which conveyances were not acquired.

One such case would be a strip passing through a so-called 20 acre parcel where one owner had acquired the North 12 acres and later the other owner had acquired the South 8 acres, but actually there was an excess of about 2/3 of an acre which belonged to neither one of these owners by conveyance and could not be prorated. This would leave a sliver between, possibly in the original owner of the whole parcel, possibly in the status of after-acquired title by the buyer of the South 8 acres, or, under prescriptive right by use of either or both owners and claimed by either/or both sides, in which case it would then require a quiet title action. Consider what would happen to this strip if it were on the diagonal through this parcel with a tie only at one end. See Figures 88 and 89. The description based on record might read:

> "Thence from a point on the south line of said west half of the southeast quarter of the southeast quarter of Section 26 distant westerly 228 feet from the southeast corner of said west half, North 17° 38′ West 1385.1 feet to the north line of said west half."

However, with the position of the north line of this parcel actually 1364 feet instead of 1320 feet north of the south line of the section, the line of right of way which must hold the bearing (since there was no lateral tie given on the north line) finds itself in foreign property on the west, thus requiring an additional conveyance or a revision of the right of way. Sufficient preliminary field survey work would reveal such a problem and save subsequent embarrassment.

Another similar example is shown in Figure 90 but with a different problem which results from there being a tie along the north line as well as along the south line. According to information obtained from the record, the scrivener began the center line of the easement "at a point on the south line of said lot, 123 feet West of the southeast corner; thence North 15° 30′ 00″ West to the north line of the South 12 acres of said lot," for the southerly parcel; and began " at a point on the south line of said lot, 123 feet West of the southeast corner; thence North 15° 30′ 00″ West to the north line of the South 12 acres of said lot and the true point of beginning; thence North 15° 30′ 00″ West to a point on the north line of said lot 395.69 feet West of the northeast corner of said lot," for the northerly parcel. Insofar as the *record* was concerned, the information conformed; however, by inspection of Figure 91 showing the measured facts, it is clear that the above form of description would necessitate an angle at the north line of the south 12 acres in order to meet the tie on the north line of the lot, and it would also traverse the sliver between the two parcels.

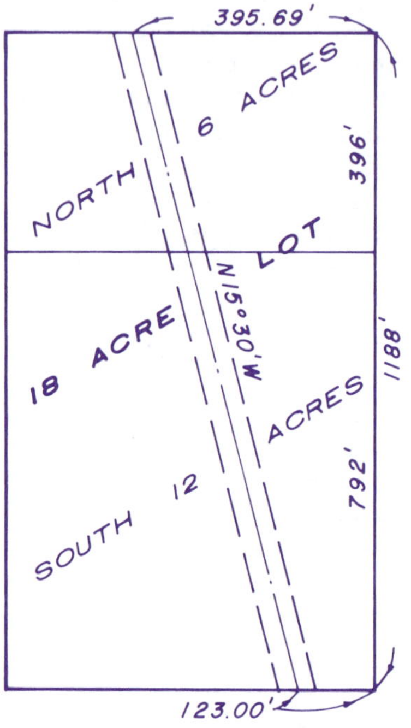

Fig. 90

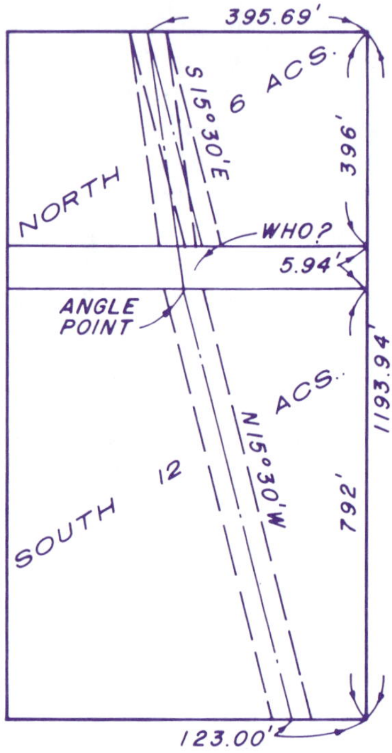

Fig. 91

If, on the other hand, the scrivener had used the reverse order for the northerly parcel and said, "beginning at a point on the north line of said lot, 395.69 feet West of the northeast corner of said lot; thence South 15° 30′ 00″ East to the southerly line of the North 6 acres," he not only would leave a gap, but even the prolongations of each strip would be offset one from the other.

Neither could the scrivener use the inclusive form of that portion of the South 12 acres included within the following described strip, and that portion of the North 6 acres, etc., etc., and use the center line based on the ties by record because the prolongation of the lot as revealed by measurements would cause an uncalled-for angle in the line and traversing over an intervening segment of land, the ownership of which was not known. So, which line should be used, and when is it possible, or is it possible, to use record information to accomplish the goal when actual measurements to the contrary are not minor but major differences from the record?

In this case at hand, each owner had bought his respective number of acres and had lived in the firm belief that that was exactly what he had, and each having planted oranges and each having used the open area along their adjoining boundary as a common turn-around for their cultivation equipment, they therefore had no finite line of demarcation between them.

When the scrivener was confronted with the facts of the excess, he queried upon which basis to set up the parcels for the right of way acquisition:

1. Prorate the acreage?
2. Determine which one was sold first and then give the excess to the other?
3. Keep them separated and find out who owned the strip in the middle?
4. Take the easement straight through and let someone else worry about the gap between?

If the easement were parallel with the side lines, the matter of alignment would be simple, but with the diagonal requirement, the gap must be considered with a possible realignment.

This is the type of situation where the title company or the title attorney called upon in advance can render a decision and cover the interest, or interests, in the strip for the protection of the acquirer.

To understand the situation clearly enough to write an unequivocal description, there should be full comprehension of the relationship between the following four elements:

1. record information from the document,
2. the strict survey of the record per se,
3. the relationship of monuments, and
4. the juxtaposition of any physical occupation.

Although it is intended that these items coincide, many times they do not.

The map should reflect these conditions, not only for the benefit of the right of way man, but also for the grantor involved and the contractor doing the improvement work. One unaccustomed to survey description work could easily be unaware of the fact that a monument may not, in truth, be the corner of his property.

Aside from insufficient research prior to writing a description, another reason why the desired coincidence is often lacking is because the scrivener who wrote the description did not include ties to physical monuments, old or new, nor ties to record monuments, such as right of ways, adjoining owners, etc. On the other hand, this does not mean the description should read from the northwest corner of Jones to the southwest corner of Smith, to an old 2″ x 4″ post, to the west end of White's land, etc. without mathematical information either. The two types of information should be properly correlated so as to give definiteness.

When the office man is planning the right of way and he finds that the reconnaissance surveyor shows in his field notes distances, angles and monuments at variance with record information, the creator of the easement-to-be needs to superimpose one on the other and then analyze the the findings to determine (1) what *must* hold, (2) what *may* give, and (3) what adjustments may be required. It may be even necessary to get additional field or record information or both in order to make the decision.

Matters which are already on record must hold in their proper delineation. The lines which may give are controlled by ties or calls to other things.

The metes and bounds type for individual parcels is desirable because it pin-points the area to be acquired from each grantor which is easier for him to comprehend and easier for the negotiator to discuss and also pleases the county assessor for tax separations. The preliminary survey work which enables this type of description to be properly developed must be sufficiently comprehensive to give the scrivener good technical selectivity in correlation with the record.

Descriptive Forms

The most common form of describing a strip of land is by use of a center line control such as:

> "A strip of land 50 feet wide, the center line of which is described as follows:

or

> "An easement for road purposes 40 feet wide lying 20 feet on each side of the following described line:"

Either because of physical circumstances or by lines of ownership, a line, other than the center line, sometimes called the survey line, may be used to control the strip such as:

> "A strip of land 75 feet wide lying 25 feet north and 50 feet south of the following described line:"

or if the strip follows around a hill or for some other reason changes direction, the wording might be thus:

> "A strip of land 75 feet wide lying 25 feet northerly, northwesterly and westerly and 50 feet southerly, southeasterly and easterly of the following described line:"

A third way is by a line completely offset on one side of the strip using the form:

> "A strip of land 46 feet wide, the southwesterly and southerly line of which is parallel and concentric with and 15 feet northeasterly and northerly of the following described line:"

There are situations where the terrain is such that extra widths are needed to accommodate cuts and fills on slope land. Very often in addition, areas of various sizes are needed as temporary easements for construction. The drawing in Figure 92 shows such a combination of permanent, slope and temporary easements divided among their respective ownership parcels.

For additional segments, with parcels of land under single ownership, you can use one of the strip type descriptions mentioned above followed by a form such as:

> "Together with a temporary easement for construction purposes over a strip of land 40 feet wide, the northerly line of which is the southerly line of the hereinabove described strip bounded on the west by the radial line passing through the easterly end of that curve cited as having a radius of 800 feet and length of 367.43 feet and bounded on the east by a line which bears S 2° 41′ 32″ W from the easterly terminus of the above-described center line."

There should be a clear statement somewhere in the document of the termination date for that temporary easement.

Additional areas for slope easements can be handled in the same way if the required shape will permit, but so often the configuration is such that an entirely distinct metes and bounds description is necessary.

Where land is held in smaller areas as shown in Figure 92, there must be even more parcels for the various conditions. The numbers shown represent parcels from different owners for the permanent easement. The addition of the letter "S" indicates an added area for slope easement and the addition of the letter "T" indicates those areas needed temporarily for construction purposes.

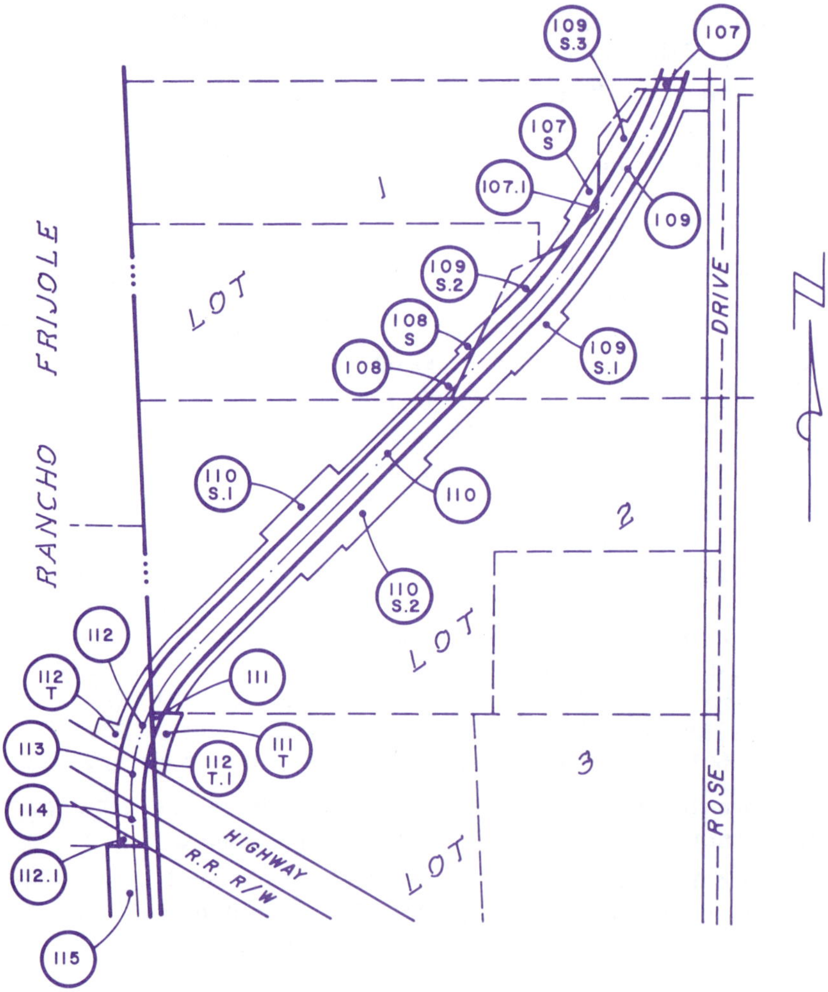

Fig. 92

Another example of parcel taking is illustrated in Figure 93 showing the parcel for temporary use designated with a decimal (2025.1). The descriptions for most of these parcels are simply "The southerly ___ feet of Lot Number ___ etc. . . . . ." or "The northerly ___ feet of Lot Number ___ etc . . . . ." Of course, the irregular shape taken in Lot 1 is described by metes and bounds.

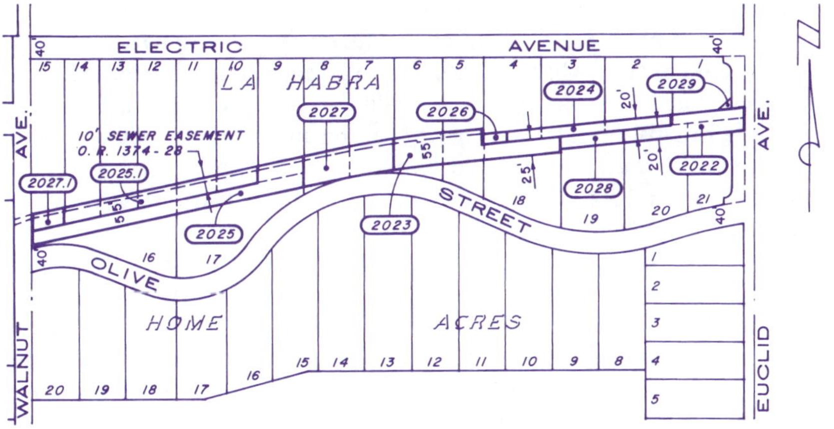

Fig. 93

Analyzing these plats delineating the parcels, one could properly ask, "Why not draft maps with sufficient detail to permit conveyance of the parcels by reference to the map?" There is no reason why such a map cannot be created and used in the same manner as a subdivision map. Preparation of the documents would certainly be much simpler by reference to the map rather than all the strip and/or metes and bounds descriptions. On the other hand, considering the chronological order of procedures in acquiring R/W parcels, one realizes that it is not always feasible to prepare a complete map ahead of time; however, the use of easement parcel maps "attached hereto and made a part hereof" with each document of conveyance not only illustrates the description but it assists the negotiator in his dealings with the owner and makes it easier for others who subsequently have occasion to use such a document.

Stationing

Some survey work is developed on a station-to-station basis; that is, the line (center line of R/W, center line of construction, arbitrary survey line, property offset line or whatever) is measured from 0 + 00 point of beginning and everything along the way such as ditches, pole lines, pipe lines, culverts, beginning and end of curves, railroads, streets, etc., is assigned a station according to its horizontal linear measurement from 0+ 00. Obversely, any distance between objects is equal to the algebraic difference between their respective stations.

When this type of line intersects another line with stationing, an equation is expressed at that point relating one to the other. If there is partial realignment, for instance, in the size of a curve which changes the position of the BC and EC, then an equation must be established at one or both of those points in order to maintain a mathematical relationship. Whereas these points of equating are apparent when looking at a map or construction plans, it requires a very clear explanation in words to maintain correct interpretation. Any recitals of equations in a description should carry specific reference to the direction of the line (and the particular line if two are intersecting) for each station given; for example, ". . . . thence S 14° 32′ W to Station 32 + 64.76 which equals (or, being the same as) Station 33 + 12.89 going southerly (or, bearing S 15° 54′ W) on the realignment of said center line, being also the center line intersection of an alley at Station 7 + 60.72 going east . . . ."

Multiple Parcels

On occasion, there may be a situation which requires that branch strip easements be attached to the main easement. Generally speaking, multiple parcels under one caption are not recommended, but if *all* of the area encumbered by the description is included in the caption and the respective pieces are properly tied to the main strip, then it may pass scrutiny well enough to be acceptable. The following example is a suggested form:

> An easement 20 feet wide over that portion of Lot C in Tract No. 436 in the County of Missoula, State of Arizona as per map recorded in Book 8, Page 4, of Maps in the office of the County Recorder of said County and those portions of Farm Lots 7 and 8 in the Butler Tract in said County as per map recorded in Book 2, Page 13, of said Maps and that portion of the southwest quarter of Section 10, T 2 N, R 4 W of the Gila and Salt River Meridian in said County, according to the official plat of said land filed in the District Land Office, the center line of which is described as follows:

Beginning at the west quarter corner of said Section 10, being the original stone marked ¼; thence S 00° 54′ W 389.42 feet on the west line of said section, being also the center line of Taco Road; thence S 89° 06′ E 20.00 feet to the True Point of Beginning on the east line of said Road; thence S 51° 17′ E 364.00 feet to Point A; thence S 51° 17′ E 189.68 feet; thence S 89° 06′ E 158.00 feet to Point B; thence S 89° 06′ E 147.00 feet to Point C; thence S 89° 06′ E 206.00 feet; thence N 58° 32′ E 458.62 feet, together with:

Parcel A: A 10 foot easement the center line of which begins at the hereinabove described Point A; thence S 89° 06′ E 289.0 feet.

Parcel B: A 6 foot easement the center line of which begins at the hereinabove described Point B; thence S 00° 54′ W 100.0 feet.

Parcel C: An 8 foot easement the center line of which begins at the hereinabove described Point C; thence N 00° 54′ E 109.0 feet.

The side lines of said 20 foot easement to be extended or shortened to meet at angle points and to terminate at the said east line of Taco Road. See figure 94.

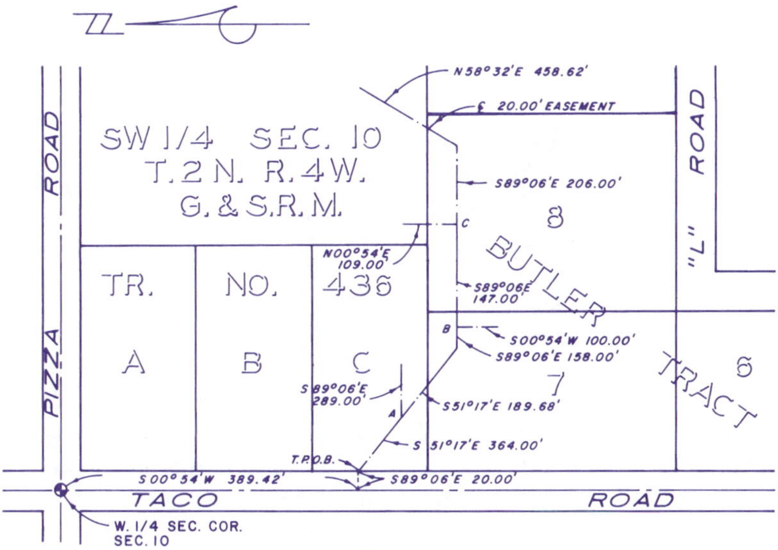

Fig. 94

Angular Termination

In terminating a strip description at a property line, caution must be used so as to neither cut off necessary parts nor add unnecessary pieces. Referring to Figure 95a, a center line description stopping at Point A with the usual form of "the side lines of said strip to be shortened or lengthened to terminate at the south-westerly line of the land described in deed to John Jones recorded in Book 349, Page 58 of Deeds." would continue the northerly line of the strip *on a curve* to that property line and thus unintentionally create the sliver parcel *Z*. The way to eliminate this problem is to extend the description beyond point A to a point B in the following tangent line and then add (instead of terminating at Jones' southwest line):

> "EXCEPT that portion thereof lying southwesterly of the south-westerly line of the land described in deed to John Jones recorded in Book 349, Page 58 of Deeds."

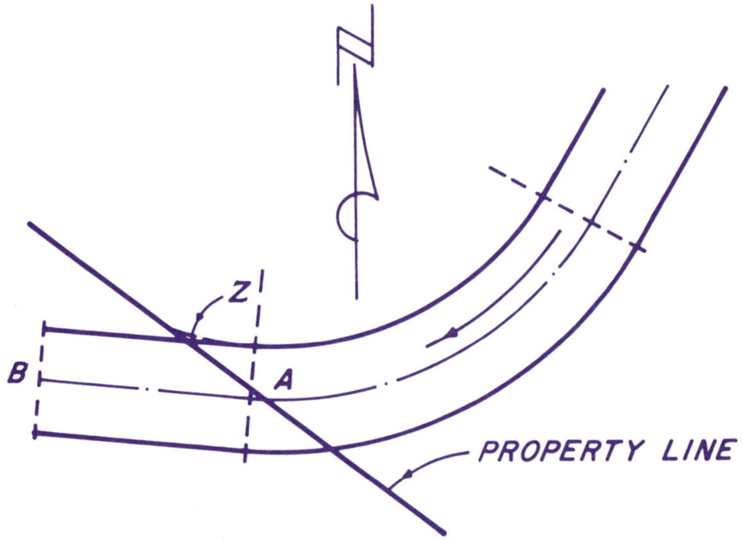

Fig. 95a

Figure 95b illustrates two problems: the one created by the description reading from northwest to southeast and stopping at the property line omits the desired triangular parcel *l* and the one created by the description reading from east to west creates the undesired triangular parcel *m*. Again, the way to omit the problem is to carry the

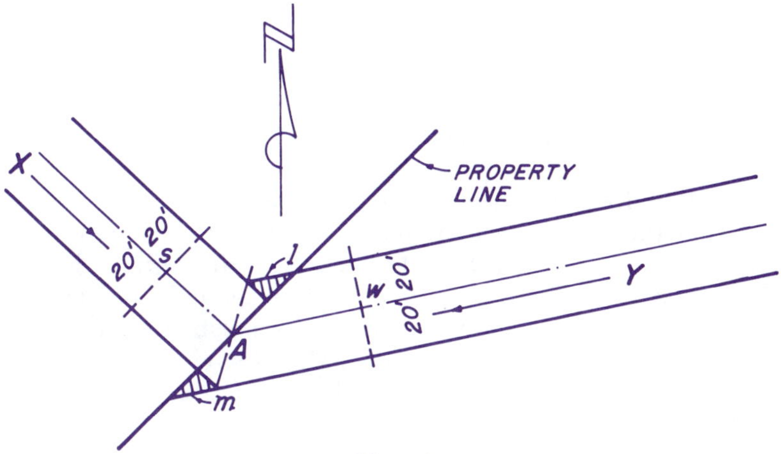

Fig. 95b

description beyond the property line, say to point S going west and northwest, or to W going southeast and east, and in each case, follow the end of the description with:

> "EXCEPT that portion thereof lying northwest (in the first case) of the southeast line of the land described in deed to . . . . ."

or

> "EXCEPT that portion thereof lying southeast (in the second case) of the southeast line of the land described in deed to . . . . ."

Other Types of Easements

Besides the commonly recognized easements discussed above, there are other servitudes on land classified by special characteristics. Some are not thought of as such, but because they are a burden upon the land, that fact places them in the category under discussion, even though under different names.

A LICENSE gives a personal privilege, unassignable and terminable at will, to do something on another's land and which contains no interest in that land and which is not required to be created by a conveyance. It does not pass to the heirs of the licensee and does not give third parties a right to sue for interference with its use. An example is where an owner gives someone a right to park on his front lawn to view a parade.

In the instance where a licensee expends labor and/or capital in the use of his license to such an extent that it would be inequitable to accept a termination of the license, it has been held that the privilege becomes in essence an easement. In one case it was held that because of improvements made by the licensee, the license became irrevocable.[94]

A PROFIT, like an easement, is described as a permissible servitude upon the land; it is the right to take something from the land of another. In California, the Code does not differentiate between Easements and Profits; the rules stated are equally applicable to both. The Civil Code Sections 801 and 802 describe *permissible* servitudes upon the land which include both easements and profits. In Georgia, it states that as a general rule, to constitute an easement, there must be both a dominant and servient estate.[95] The right to profits a prendre is in its nature corporeal, and is capable of livery, while easements are not and may exist independently without connection with or being appendant to other property. (Id. 279, 1-C) An example of PROFIT is taking a crop of wheat from another's land or removing coal from another's land. In the latter case, ample support for the surface of that land must be provided as a part of the responsibility attached to the right to take the profit.

There are certain NATURAL RIGHTS incidental to the ownership of land which attach to and pass with the conveyance of the land, even though not created by contract or conveyance. They are not in fact easements because they are not rights in another's land. For example: The right to have surface waters flow over one owner's land onto the adjoining land; the right to receive lateral support from adjoining land for the preservation of the upper soil in its natural state.

NEGATIVE easement is another expression for equitable servitude. It is exemplified by the preservation of light and air space as previously mentioned or by tract restrictions thereby burdening each lot in favor of each other lot.

Other rights, servitudes, or burdens which may be attached to land as appurtenances or incidents include:

a. The right of having water flow undiminished and undisturbed.
b. The right of flooding land.
c. The right to take minerals, water, wood, fish, game, etc.

[94]Hammon v. Mustard 257 CA 2d. 384, 64 CR829.
[95]152 S.E.575.

d. The right of pasture.

e. The right to use a wall as a party wall.

For most easements that require a land description, the planimetric form is suitable. In a few instances, a three-dimensional description will be required. In the case of "rights" mentioned above, the description of the grantor's land will often suffice.

PROBLEMS

1. Explain the difference between a "gross" easement and an "appurtenant" easement. Give an example of each.
2. Explain "overburdening" as applied to easements.
3. Name three types of control for "strip" easements.
4. Explain the difference between a "negative" easement and a "positive" easement. Give an example of each.
5. How many easements can be permitted over the same strip of land?
6. From the facts shown in Figure 87, write descriptions for the following parcels with appropriate easements and reservations.
   Parcel U, as a whole unit without any subdivision within it.
   Parcel V.
   These parcels are being created out of a Lot in a Tract shown on a recorded map; therefore, write your own caption for the Lot. Do not refer to any of the Parcels U, V, W or X in your descriptions because they do not exist on record.
   Do not use a metes and bounds type of description — use the short form.
7. In consideration of the fact that writing descriptions using "stationing" requires explicit wording to substitute for not always having the plan and profile before you to see the facts, what is *one* of the specific requirements for making certain that the stationing described is clear to the reader?
8. In writing a strip description that ends at a line which crosses the strip on a curve or at an angle point, what wording should you use to prevent the inclusion of extra unnecessary parts and/or the exclusion of necessary parts?

— NOTES —

Chapter 13
Descriptions of Section Land

Government Created

The subdivision of the public lands of the United States of America is surveyed under, and known as, the United States System of Rectangular Surveys.

Before getting into this matter very deeply, there are several items that need to be clarified in order to understand the reason why some commonly expressed concepts of section land are not acceptable. It is not because of the rules set forth in the Manuals of INSTRUCTIONS FOR THE SURVEY OF PUBLIC LANDS in the UNITED STATES; rather, it is the lack of understanding of what actually happens in the field. You are familiar with the theoretical layout of a township having six rows of six sections and that presumably each section is one mile square with perhaps a little less than one mile east and west in the tier of sections along the west side of the townships and likewise a little less (or perhaps a little more) than one mile north and south in the row of sections along the north side of the township; the variation of which was planned to accommodate the convergence of the meridians.

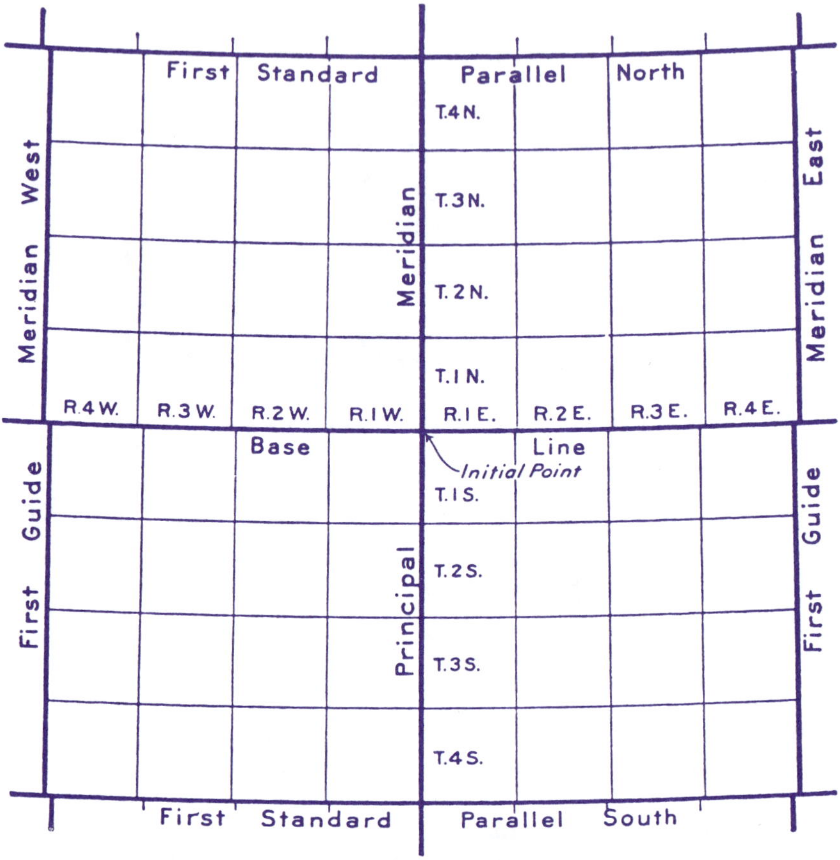

Standard parallels and township boundaries are given precedence over subdivisional lines.

The great majority of the township plats submitted, and approved, in every area of government surveyed lands for the first time conformed to the above format. It was when new township plats were made from resurveys that discrepancies were clearly illustrated showing the results of the earlier field work. Some of the grotesque shapes and sizes of sections are difficult to believe, but for the facts shown on the plats and in the field notes. If you have not already seen some of these, no doubt you can find examples in the files of your particular District Land Office.

At this point, we should analyze and understand the instructions used by the government surveyors as presented by the General Land Office, now the Bureau of Land Management. The following is reproduction of a pertinent part of a notice* attached to the inside front cover of a June 1, 1909, Edition of the GLO pamphlet circular, "Restoration of Lost or Obliterated Corners and Subdivisions of Sections."

THE MANUAL OF SURVEYING INSTRUCTIONS.

Department of the Interior,
General Land Office,
Washington, D.C.,July 1, 1909

* * *

County surveyors, attorneys and other local people sometimes are under the impression that in the Manual are found the proper means to guide their action in making private resurveys, restoring lost lines and corners of old surveys, and subdividing sections, where parties are in conflict. They should be advised that the instructions do not apply to such secondary work, but only to the present official and original surveys, those about to be executed under authority of the Government, and under the provisions of which all original public land surveys have been made by the United States since the date (1902) of the Manual.

The advice needed in restoring lost lines, etc., is offered by this office in a pamphlet circular dealing with all such cases in general terms, and entitled "Restoration of Lost or Obliterated Corners, and Subdivision of Sections", (revision of June 1, 1909) which is mailed free by this office to all applicants.

* * *

A revision of the Manual, extending the tables for another decade, will be made in 1910.

Commissioner.

In other words and briefly:

1. The Manuals of Instructions were intended for the use of those in Government service.
2. The Corner Restoration Circulars were prepared for the private sector — county surveyors, private surveyors and engineers, attorneys and others.

*Courtesy of Carlisle Madson, L.S., Hopkins, Minnesota.

This does not mean that one should ignore the Manuals in reestablishing Township corners and lines and/or Section corners and lines; one should learn and understand the methods promulgated and used by the government surveyors at the time they executed their contracts for those surveys. This, combined with the information in the pamphlet on "Restoration of Lost or Obliterated Corners and Subdivisions of Section," along with all the physical evidence available, are the tools with which you reconstruct government-created corners and lines.

Remember that the date of the Manual or Pamphlet applies to the work done following that data and prior to the next issue.

Another aspect to consider in reconstructing lines of title in government section land is that all prior private claims, such as military grants, mineral claims, homesteads, private grants, etc., were presumably reestablished by the government surveyor before setting section corners in that vicinity. The remaining fractional parts of the section would then be lotted according to instructions.

Now, what does this have to do with descriptions? Too often the assumption is accepted that a section is just one mile square and that therefore the quarter section is 2640 feet by 2640 feet and the subdivision of that is 1320 feet, 660 feet, 330 feet, etc. It is an extremely rare case when a section line between correct monuments is measured exactly 5280 feet or 2640 feet to the quarter section monument. More commonly, there is found a difference in measurement all the way from the best part of a foot up to many feet and sometimes more. In fact, one particular Section 6 is a mile and one half east-west, while in another area, a certain Section 3 is over a mile and one half north-south. For this reason, the layman should be apprised of the fact that he should not rely on subdividing the section on paper because in all probability a resurvey on the ground in conformance with the rules will disclose any number of differences.

One reason for the likelihood of discrepencies is understood when you realize the amount of error of closure that has been permitted by the Manuals. In the first one of 1855, it states, "The east and west section lines, except those terminating on the west boundary of the township, are to be within one hundred links of eighty chains in length; and the north and south boundaries of any one section, except in the extreme western tier, are to be within one hundred links of equal length." In the latest Manual of 1973, it states, "The 'limit of closure' set for public land surveys may now be expressed by the fraction 1/905, provided that the limit of closure in neither latitude nor departure exceeds 1/1280 .... An accumulative error of 6¼ links per mile of perimeter, in either latitude or departure, will not be exceeded in an acceptable survey."

"The latitudes and departures of a normal section shall each close within 25 links, of a normal range or tier of sections within 88 links, and of a normal township within 150 links."

Compare this with the private surveyor's closure of 1:5,000 to 1:10,000 in many cases, and, in some areas, even greater accuracies, and the national Geodetic Control Survey accuracy requirements from Third Order (2) 1:5,000 min. to Base Line 1:1,000,000 min. This is not to say that the survey of public lands should adhere to a greater accuracy; this is to point up the fact that you should be fully aware of these differences and not expect a description of "the west half of the southwest quarter of Section 12," to necessarily yield a 1320 foot by 2640 foot parcel of land.

Furthermore, the rules still apply even if the section be distorted. For example, Figure 96 shows the configuration resulting from found monuments; still the division into halves and quarters must follow the rule of joining opposite quarter corners with straight lines. Many examples of oddities in sections can be cited but the point to remember is, follow the rules in every way possible.

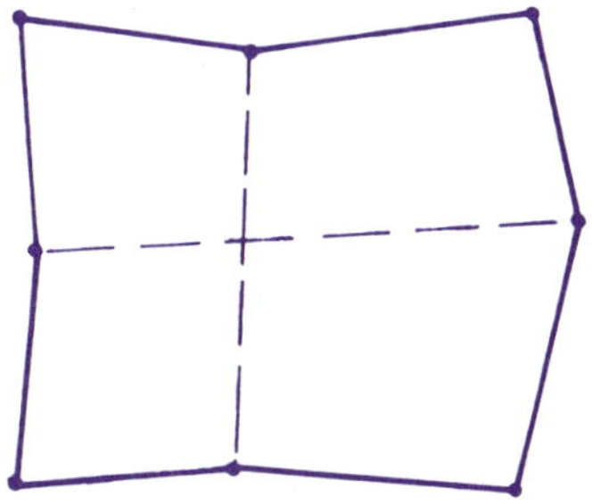

Fig. 96

Incidentally, there also exist odd situations in townships. In those areas where one Meridian bounds another, the townships are very often fractional. Obversely, there is one township that has sections numbered up to 46!

A different reason for some of the bigger discrepancies is that separated areas have been surveyed without having first established the entire framework of township lines; consequently, there were voids containing excessive distances which were unknown until subsequent surveys filled them in.

13.6

The numerical order for Townships, ranges, sections and lots is explained in Sections 3-1 through 3-84 in the 1973 Manual. When a section does not contain the regular 640 acres, as in the west tier or north row in a township, the odd areas are assigned lot numbers. See Figure 97a.

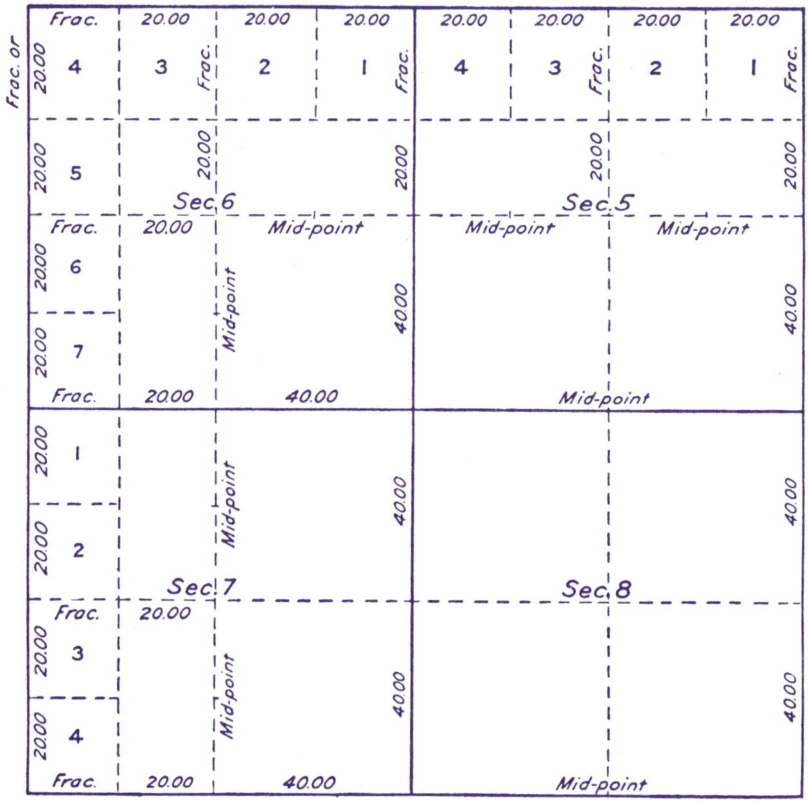

Showing normal subdivision of sections.

Fig 97a

When sections are cut up by a river, a lake, a rancho, etc., the land between the regular areas (usually down to 40 acres) and the irregular line is cut into lots. Figure 97b shows four typical examples from the Manual.

SYSTEM OF RECTANGULAR SURVEYS

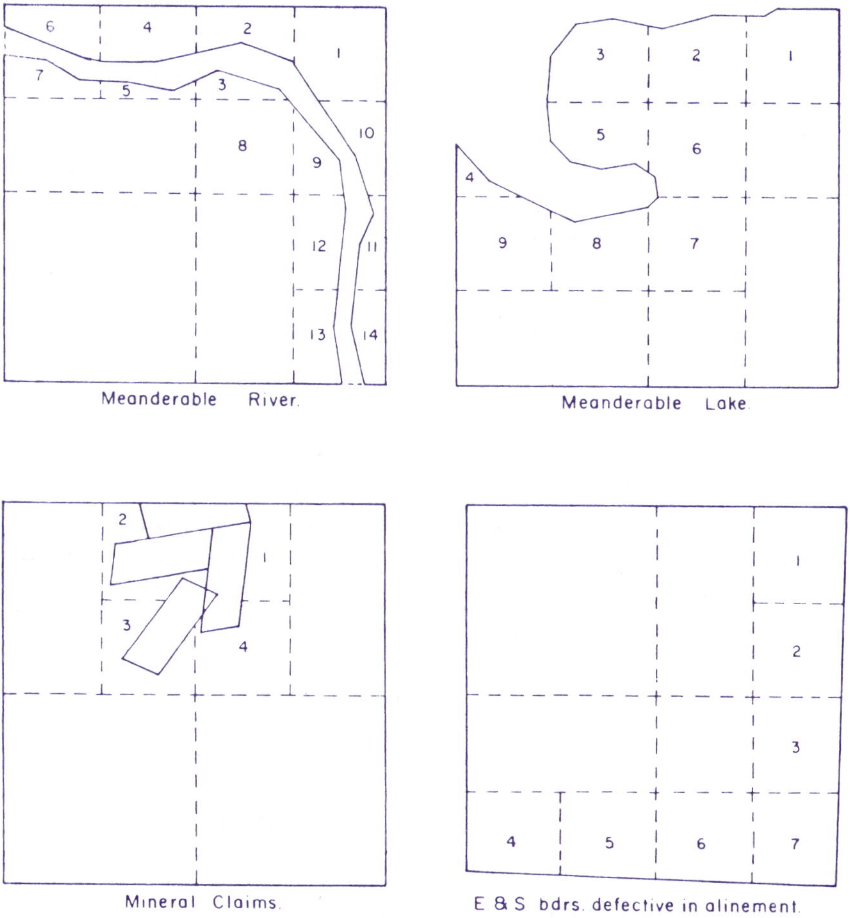

Fig. 97b — Examples of subdivision of fractional section with lots.

MANUAL OF SURVEYING INSTRUCTIONS
BLM Technical Bulletin 6

Lots in sections on an approved township plat are "whole" lots, not "fractional" lots. The very reason they exist is because the area so designated is a fractional part instead of a regular part (40 ac, 80 ac, etc.) of a section. The word "fractional" is properly applied to those sections cut off in one or more parts by reason of their irregular boundary.

The subject matter used in a section type description should be arranged in the following order:

Lot Number (if any)
Section (or part of)
Township (North or South)
Range (East or West)
Meridian
City (if any)
County (or Parish or Borough)
State
Reference to the Official Plat of said land filed in the District Land Office

A simple example of this is:

"Lot 1, Section 4, Township 3 North, Range 2 East, of the First Principal Meridian, in the County of Oshkosh, State of North Dakota, according to an official plat of said land filed in the District Land Office."

Again, as before, you may have a portion or portions of a section which may be designated specifically or "described as follows:." If it is "The southwest quarter of the northeast quarter of Section . . . ," you have a definite part of the section described and do not need "described as follows:." If you have only a part of this, the form will be thus:

"That portion of the southwest quarter of the northeast quarter of Section . . . . "

continuing with:

"described as follows:"

and then add the body of metes and bounds. Each of these forms is to accommodate the particular kind of area with which you are concerned.

Notwithstanding the rules for numbering lots in a section as set forth in the Manual, there are some approved township plats which carry duplicate numbers in the same section. In such cases, the wording must be of a very particular form in order to assure the correct conveyance of the intended part.

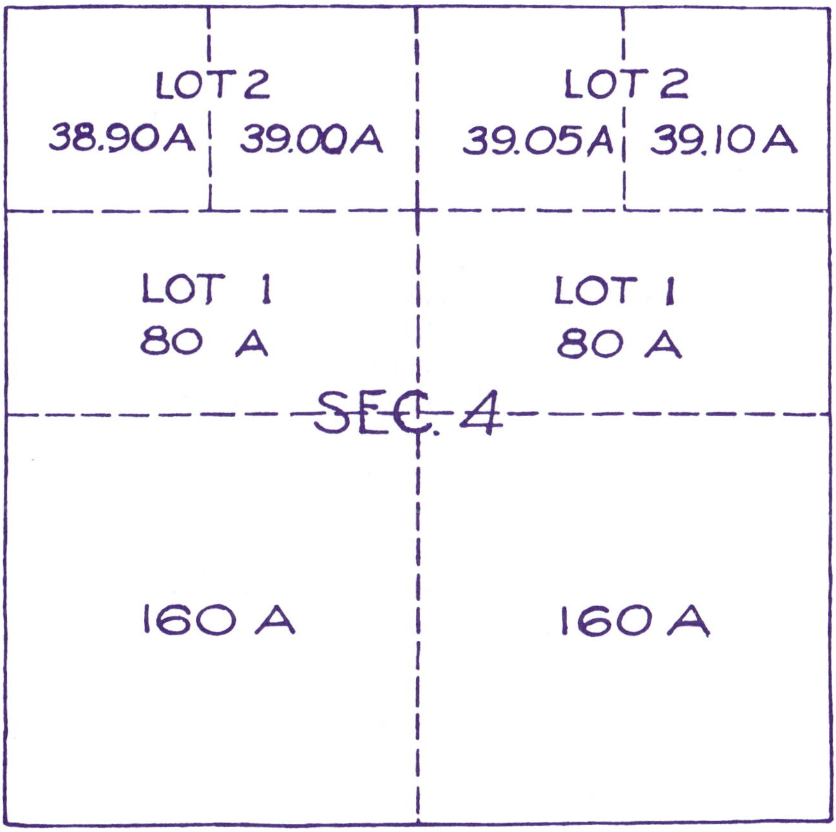

Fig. 98

Looking at Figure 98, you see two pairs of duplicate lots. Because both of the areas labeled LOT 1 are shown as containing 80 acres, which is regular for the south half of the northwest quarter and the south half of the northeast quarter, neither one was required to be assigned a lot number. Although it might be acceptable, because of the facts shown here, to use a description of "the south half of the northwest (or northeast) quarter of Section 4 . . . . " which would conform with the showing on the official plat, the lot number would be proper but it must carry a correlative designation of the section quarter in which it lies.

In regards to the other duplicate lots it is another story. Fortunately, each acreage is different and it will be the determining factor.

Your form of description for these lots should read:

"That portion of Lot 2 in the northwest (or northeast) quarter shown as having ______ Acres in Section4, T ___ ___, R ___ ___, in the ______ Meridian, in the County of ______________________, State of ______________________ according to the official plat of said land filed in the District Land Office."

Another matter concerning lots in a section subdivision is that of water boundaries. The meandering line of the lot along the river, creek or stream is, as stated before, a survey line for the determination of area, while the line of title ownership follows the high water mark, the low water mark, or the thread of the stream, according to whatever determining conditions exist. Consequently, reference to the lot itself carries whatever title the patent conveyed.

The use of "meander" lines is not confined just to water boundary areas. In the case of mountain ridges, the line following the peak elevations has been described as the boundary line of title whereas the survey traversed from one prominent point to the next one on a meander line principle. Again this was for the purpose of ascertaining a quantity of land for the conveyance.

When a description is written with good intent but without sufficient qualification, it creates a question as to the specific intent. Consider the lines shown in Figure 99 where the description in the deed read, and without a caption,

Fig. 99

"Beginning at the southwest corner of Section 11, T ___ ___, R ___ ___, of the ______ Meridian; thence North 10 chains, East 10 chains, South 10 chains, West 10 chains to beginning."

Again this apparently was on the presumption of a *regular* north-south and east-west section. When such language is used in this manner, it has been held that such terms, lacking any qualification, are construed as equivalent to, and based upon, an astronomically established north on a meridian of longitude.[96] The "east" and "west" lines are considered as straight at right angles to that meridional line.

From the drawing, the obvious problem is lack of conformity between the described lines and the section lines. If there had been a caption of "That portion of Section 11 . . . etc.," the triangle included in Section 10 would be automatically cut off and there would be the triangular gap along the south line of Section 11. If, on the other hand, the description had been written not only with a caption but with each line properly qualified, the result would have given a parallelogram conforming with the boundaries of the section. A correct form would read thus:

> "That portion of Section 11, T ___ ___, R ___ ___, of the ______ Meridian in the County of ______________________, State of ______________________, according to the official plat of said land filed in the District Land Office described as follows: Beginning at the southwest corner of said Section; thence north 10 chains along the west line of said section; thence east 10 chains parallel with the south line of said section, thence south 10 chains parallel with the west line of said section to the south line of said Section; thence west 10 chains to the point of beginning."

This wording would hold the west and south lines coincident with the section lines in spite of the fact that they differed from "true" lines as much as 8° and 6° respectively.

A short form description which would accomplish the same objective would read thus:

> The southerly 10 chains of the westerly 10 chains, measured along the west and south lines respectively, of Section 11 T ___ ___, R ___ ___, of the ______ Meridian in the County of ______________________, State of ______________________, according to the official plat of said land filed in the District Land Office."

Sometimes a parenthetical phrase is inserted and whether it is in fact parallel, supplementary or predominate depends on the form of its content. This kind of supplemental information can be helpful but, by wrong use, it can also be confusing or even nullifying in its effect.

[96]Richfield Oil Corp. Crawford, 39 C 2d. 729, 249 P 2d. 600.

A simple type of parenthetical description would read thus: —

The west half of the northeast quarter of the northeast quarter of the northeast quarter of section 10, T ___ etc. described as follows:

> Beginning at a point on the north line of said section South 89° 42′ 55″ West 328.78 feet from the northeast corner of said section; thence South 89° 42′ 55″ West 328.78 feet; thence South 0° 03′ 24″ West 662.18 feet; thence North 89° 36′ 48″ East 329.02 feet; thence North 0° 01′ 36″ East 661.96 feet to the point of beginning.

This indicates that a refined survey of the section had been made and that the bearings and distances cited reflect the true ground conditions. These would hold over the fractional reference because they are more definite.

Some descriptions of section land not only tie to and along the section (or sub-section) line but also to adjoining owner's land. This can introduce the problem of complicity from excess ties if the adjoiner's line is not coincident with the section (or sub-section) line. For instance, if the description of the land on the east side of the sub-section line shown in Figure 100 ties also to "the east line of Tom Brown's land on the west" it can raise the question of which one in fact will control. Extensive research back through the chain of title may then be necessary to determine the validity of references.

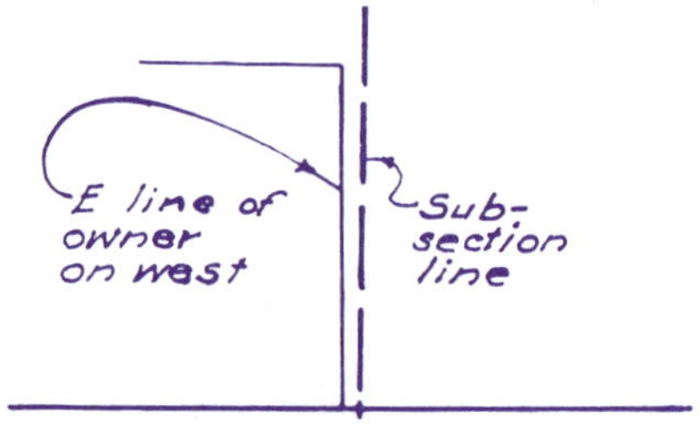

Fig. 100

If the obverse were the case, that is, if "the east line of Tom Brown's land on the west" was in fact *east* of the sub-section line, you should add an EXCEPTION if it is a good fee title or, if there is a question, show the instant property "Subject to the interest in the land described in deed to Tom Brown recorded in Book ___, Page ___ of Deeds included within the above described land."

When title is held within a quarter section or quarter-quarter section or some other definitive part of a section, there is no need of making a tie to the adjoiner's land by the same analysis that when you refer to all or part of a lot in a subdivision, you do not tie to John Jones' land in the adjoining lot because the lots have been established by survey and your reference to the lot according to the map is sufficient. Likewise, when the sub-section lines are established by survey and/or the adjoiner's deed ties to the same sub-section line, that is sufficient.

The need for descriptive ties to the adjoining land is paramount in describing and following lines not established by rules or not shown on previous maps and for which there is no other record information.

Private Survey Created

There are a few private land holdings that have been subdivided on a section type basis with townships and ranges. There was no requirement that the surveys and maps of these areas follow government rules and they did not, even though the maps indicate an approach to such similarity.

The question arises as to what form of description should be used. First of all, go to the map (recorded or not) and read the caption on it for source of title and study the pattern used in developing the sections; are they laid out in townships and ranges; are they regular or irregular; do they contain lots; are the sub-section lines delineated with measurements?

Next, find the correct reference to use for the map. If it is recorded, there will be a book and page to use. If not recorded, you need some way to identify it and disclose its whereabouts for further references.

Unless there is a radical departure from the governmental plan, you can use the same form of description insofar as it applies to the various parts. The most valuable requisite is that you have a bona fide map reference, and use the *source* of title in your caption. Following are three examples of descriptions for private-land-surveyed-as-sections. The first one was recorded in the same books as subdivision maps. The second one was filed as a public record of a private survey in Records of Surveys. The third one was the result of a court decree and so reflects that fact.

* * *

Section ___, Township ___ South, Range ___ West, in the Rancho ______________________, City of ____________________________, County of ____, State of ______________________, as shown on a map recorded in book 51, page ___ of Miscellaneous Maps, in the Office of the County Recorder of said County.

* * *

Lot ___ in Section ___, Township 4 South, Range 11 West, in the City of ____________________________________,County of ________________________, State of ________________________, as shown on a map of a Resurvey of the J. W. Bixby and Co's Subdivision of a part of the Rancho Los Alamitos, filed in book 2 page 43 of Record of Surveys in the Office of the County Recorder of said County.

* * *

Section ___, Township ___ South, Range ___ West, in the Rancho Los Alamitos, City of ______________________________, County of ________________________, State of ________________________, as shown on Map No. 2 attached to the Final Decree of Partition of said Rancho, a certified copy of which was recorded February 2, 1891 in book 14, page 31 of Deeds, in the Office of the County Recorder of said County.

* * *

Notice the reference to origin of title. It just so happens that each of these came out of Rancho grant land.

PROBLEMS

1. You may have "fractional" sections; do you also have "fractional" lots in a section?
2. In the case of duplicate numbered lots in the same section, how do you discern one from the other in a description?
3. Is the "meander line" of a lot the limit of ownership?
4. If a boundary of the property being described is a lot line, either one shown on an official plat of a township or on a recorded tract map, or a section line shown on an official plat or a sub-section line developed according to the rules, is it desirable to tie to the adjoiner's land also?
5. What is the essential control for the recovery, reestablishment or subdivision of government type sections created by private survey on privately-owned lands?

— NOTES —

Chapter 14
Special Shape Descriptions

There are increasing demands for different formats to describe certain specific shapes in a one plane area and volumetric forms in the three dimensions of the two plane concept, although they are not as common as the run-of-the-mill descriptions.

Planimetric Descriptions — Special Shapes

The description of a hexagon or octagon or any multi-equal-sided form can be accomplished without following by metes and bounds around the perimeter. The principle involved is merely to declare the pattern of the form and then relate the bearing and length of one of its sides to a control line or point, being certain that the base of bearings is included. Follow Figure 101 for one example:

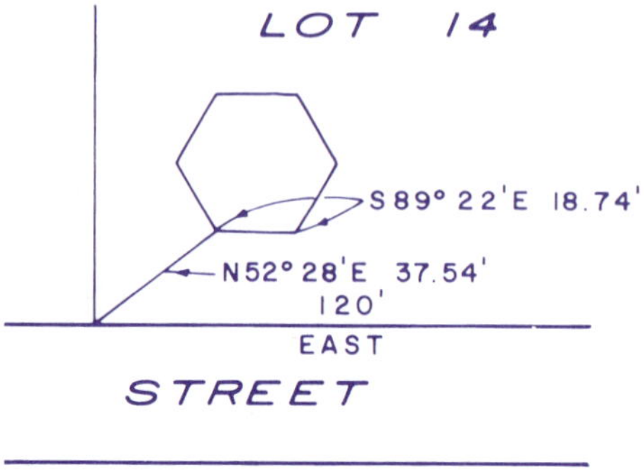

Fig. 101

A hexagon in Lot 14 in Tract No. __________ in the City of ________________, County of ____________________, State of ________________________, as per map recorded in Book ___, Page ___, of Maps in the office of the County Recorder, the southerly line of which is described as follows: Beginning at a point N 52° 28′ E 37.54 feet from the southwest corner of said Lot; thence S 89° 22′ E 18.74 feet.

Of course, the bearings in this case are based on those shown on the map to which reference is made in the caption and the remaining sides of the hexagon can be established by calculations or by drafting.

To create a cul-de-sac at the end of a street butted against the end lots as shown in Figure 102, use a description reading thusly:

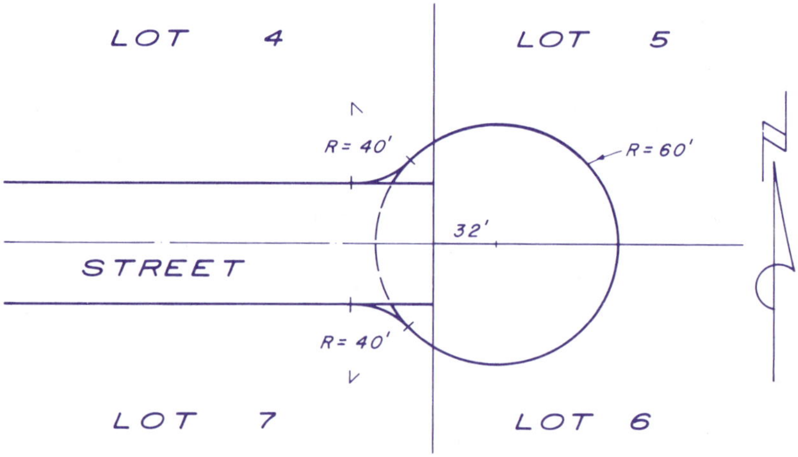

Fig. 102

Those portions of Lots 4, 5, 6 and 7 in Tract No. ___, in the City of ____________, County of ____________, State of ____________, as per map recorded in Book ___, Page ___, of Maps in the office of the County Recorder of said County included within the circumference of a circle having a radius of 60 feet, the center of said circle being on the line common to said Lots 5 and 6 east 32.00 feet measured along said line from the southwest corner of said Lot 5.

Together with that portion of said Lot 4 lying southerly of a curve concave to the north having a radius of 40 feet which curve is tangent to both the aforedescribed curve having a radius of 60 feet and the north line of N Street shown on said map.

And together with that portion of said Lot 7 lying northerly of a curve concave to the south having a radius of 40 feet which curve is tangent to both the aforedescribed curve having a radius of 60 feet and the south line of N Street shown on said map.

If you have all the calculations for these curves, you could incorporate them into a metes and bounds description which would be approximately the same length as this. Either way will pass good title.

On occasion, it is merely a matter of rounding the corner of a lot in which case you can use the following form, whether the existing corner is acute or obtuse.

That portion of Lot ___, in Tract No. ______, in the City of ____________, County of ____________, State of ____________, as per map recorded in Book ___, Page ___, of Maps in the office of the County Recorder of said County lying between the (easterly) and (southerly) lines of said lot and a curve concave to the (northwest) having a radius of (35) feet which curve is tangent to said (easterly) and (southerly) lines of the lot.

This is perfectly good for the conveyance of title because the curve can fit only one position. The balance of mathematical information, if it is desired, can be obtained by calculations from the record map or by field survey. The same format can also be adapted to fitting a curve tangent to curved lot lines.

Tri-dimensional Descriptions

Turning now from the flat areas to vertical enclosures, we find the old adage, "an individual owns his land from the center of the earth to the heavens above," is being changed; the expanses of heaven and earth are being forced into much smaller confines by pressure of the expanding volume of the requirements of our economic growth compacting the volume of usable space. It is being molded into various configurations to accommodate such items as tunnels, bridges, buildings, airport corridors known as avigation easements, condominiums, etc. Some existing examples of aerial rights for the use of space over railroads or highways are such as the Prudential Building, the famed Merchandise Mart and the cylindrical twin Marina City Towers in Chicago, the PANAM Building and Kratter Corporation's Bridge Apartments in New York, the Bethesda Air Rights Building in Maryland and others.

As the movement of a point develops a line, the movement of a line develops a plane and the movement of a plane develops a solid, so must the description of the horizontal plane, expanded from the point and line, be moved into the third dimension to develop the multi-sided form. The discussion on condominiums showed how a simple way to describe a many-sided space can be accomplished by reference to a map. Tunnels, bridges, apartments or the like can be delineated on a drawing with sufficient information that their volumetric form can be conveyed by proper words in the document referring to the map. An excellent example is a certain bank building which straddles a street; both the planimetric and vertical controls were delineated on a parcel map and such matters of concern in the lease were merely referred to that map as recorded. However, in spite of the desirability of using maps, there still persists the demand for worded legal descriptions in this realm.

A metes and bounds description around a multi-sided space may be accomplished by the method of a continuous direct perimetrical delineation, but, depending on the shape, it may become extremely involved and unwieldy. To illustrate a simple form, consider the matter of taking a vertical cut out of a street for the purpose of lowering the grade. The description is illustrated in Figure 103.

> "That portion of the West 30.00 feet of North Nicolas Avenue adjoining Lot 12 in Tract No. 201 in the City of ____________________, County of ____________________, State of ____________________, as per map recorded in Book 13, Page 6, Miscellaneous Maps in the office of the County Recorder of said county, bounded northerly by the easterly prolongation of the northerly line of said Lot 12 and bounded southerly by the easterly prolongation of the southerly line of said Lot 12 and included within the vertical section lying above the following described plane:

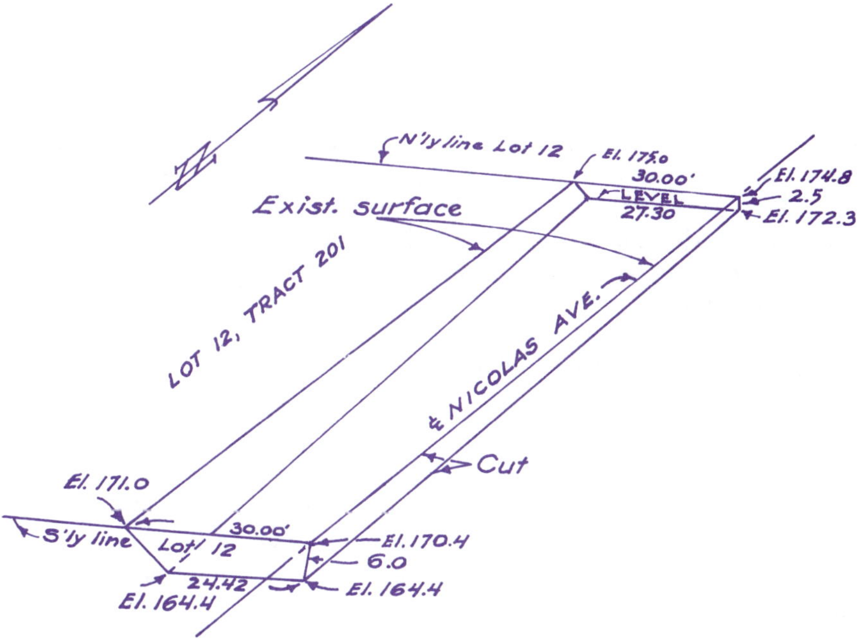

NICOLAS AVENUE LOWERING

Fig. 103

Beginning at the northeast corner of above described West 30.00 feet, the existing surface elevation of said point being Elevation 174.8 feet per city datum as shown on Plan File No. 1264 on file in the office of the City Engineer of said City; thence vertically downward 2.5 feet to Elevation 172.3, said point being the True Point of Beginning; thence, along the east line of said West 30.00 feet, Southerly and descending to Elevation 164.4 in the southeast corner of said west 30.00 feet; thence, along the southerly line of said West 30.00 feet, Westerly 24.4 feet to a point at Elevation 164.4 in a line having a downward slope of 1 foot vertical to 1 foot horizontal, measured from the existing surface elevation (Elevation 171.0) at the southeast corner of said Lot 12; thence, Northerly and ascending to a point at Elevation 172.3 in the northerly line of aforesaid West 30.00 feet, said point being in a line having a downward slope of 1 foot vertical to 1 foot horizontal, measured from the existing surface elevation (Elevation

175.0) at the northeast corner of said Lot 12; thence, along the north line of said West 30.00 feet, Easterly 27.3 feet to the True Point of Beginning.

TOGETHER WITH a longitudinal section of triangular cross section, the lower plane of which is that certain plane connecting the 1 foot vertical to 1 foot horizontal line in the southerly line of said West 30.00 feet with the 1 foot vertical to 1 foot horizontal line in the northerly line of said West 30.00 feet, as said lines are hereinabove described."

A map, or a perspective drawing as shown in Figure 103, "attached hereto and made a part hereof" is always helpful in clarifying the relationship of the different planes.

Aerial

A typical easement for a bridge is a combination of an elongated cube-like form to encompass the bridge itself plus two or more vertical cylindrical or cubical forms standing between the bridge-in-the-air and the ground for support structures. The perspective drawing in Figure 104 illustrates an example of the relationship between the physical structure and the parts of the easement.

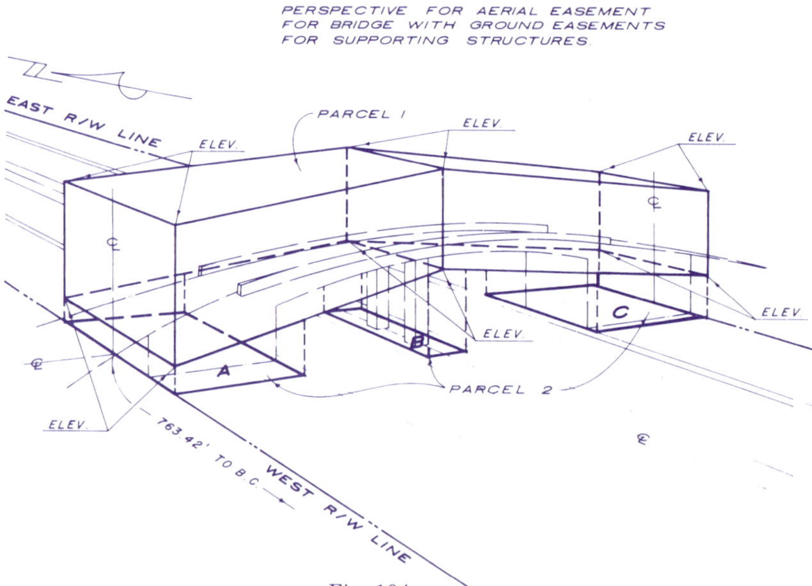

Fig. 104

In essence, the fundamental descriptions are planimetric and the elevations which control the vertical limits are set in a complementary description. Figure 105 shows the plan of the horizontal control for the example bridge easement. A suggested form would read thus:

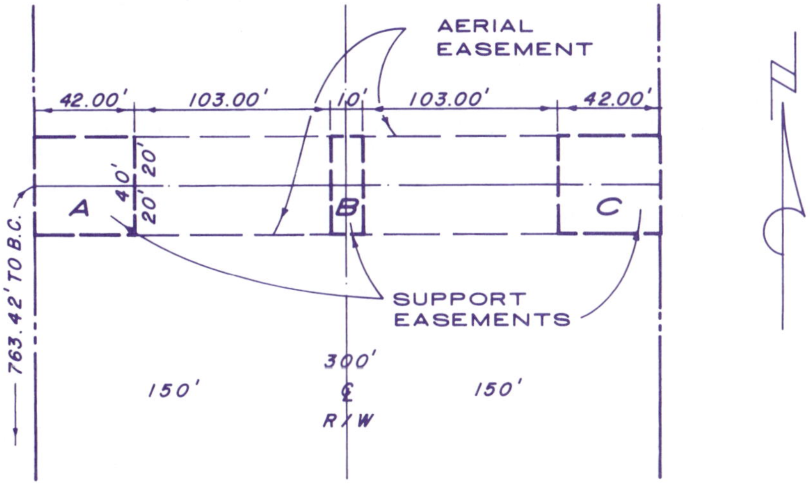

Fig. 105

PARCEL 1:

An aerial easement for road and bridge purposes 40.00 feet in width over and above that portion of Section 12, T 4 S, R 8 W of the ______ Meridian in the county of ______________, state of ______________ according to the official plat of said land filed in the District Land Office, the center line of which is described as follows:

Beginning at a point on the west line of the 300 foot right of way described in deed to the Division of Highways, state of ______________, recorded May 16, 1965 in book ___ page ___ of Official Records, distant North 763.42 feet from the southerly terminus of that certain course in said deed cited as "South 1259.62 feet to the beginning of a curve"; thence at right angles East 300.00 feet to the east line of said right of way; the bottom plane of the vertical space contained within said easement shall be (based on county surveyor's B.M. No. 362 Circuit No. 4, checked April 9, 1966, CSTB 140-A page 56) from elevations of 78.52 feet on the west line of said right of way, to elevations of 90.78 feet on the center line of said 300 foot right of way to elevations of 79.08 feet on the east line of said right of way; and the top plane of the vertical space of said easement shall be 20.00 feet above said elevations cited for the bottom plane.

PARCEL 2:

40 foot wide easements for support and maintenance of a bridge within the aerial easement described in Parcel 1 above, over and above those portions of Section 12, T 4 S, R 8 W of the ______ Meridian in the county of ______________________, state of ______________________, according to the official plat of said land filed in the District Land Office, the top plane of said easements being coincident with the bottom plane described in Parcel 1 above and the center lines of which are described as follows:

EASEMENT A:

Beginning at a point on the west line of the 300 foot right of way described in deed to the Division of Highways, state of ______________________, recorded May 16, 1965 in book ___ page ___ of Official Records, distant North 763.42 feet from the southerly terminus of that certain course in said deed cited as "South 1259.62 feet to the beginning of a curve"; thence at right angles East 42.00 feet.

EASEMENT B:

Beginning at the above described point on the west line of said 300 foot right of way; thence at right angles East 145.00 feet to the true point of beginning; thence East 10.00 feet.

EASEMENT C:

Beginning at the above described point on the west line of said 300 foot right of way; thence at right angles East 258.00 feet to the true point of beginning; thence East 42.00 feet to the East line of said right of way.

Again the attachment of a map and perspective drawing can be very enlightening. If a perspective is too difficult to obtain, even a plan and vertical cross section would be helpful.

Subterranean

Easements below the surface of the ground can be written generally with the same format as the bridge description. Elevations will still control in their respective juxtapositions. In lieu of bridge supports, there may be required vertical air supply duct, elevator, stairway or ramp easements as necessary appurtenances. For a straight segment of tunnel to be acquired from a certain property owner, one might use wording such as:

PARCEL 1:

A cylinder 30 feet in diameter through that portion of the Rancho La Puente, in the County of ______________________, State of ______________________, as shown on a map recorded in book 1, pages 43 and 44 of Patents, in the Office of the County Recorder of said County, the center line of which is described as follows:

Beginning at a point on the east line of said Rancho N 3° 22′ 36″ W 768.59 feet along said east line from its intersection with the northerly line of Foothill Road as described in deed to said county recorded November 3, 1972 in Book 1652, Page 38 of Deeds; thence S 72° 36′ 41″ W 1863.78 feet to a point on the west line of the land described in deed to William Hagensburg recorded February 9, 1972 in Book 1475, Page 24 of Deeds, said point being N 2° 46′ 12″ E 567.89 feet along said west line from the said northerly line of Foothill Road, the side lines of said cylinder to be shortened or lengthened to terminate at the above described east and west lines. The elevation of the center of the circle of the cylinder at the point of beginning to be (based on USC & GS Station Los Cerritos, 356.2 Feet,) elevation 293.8 feet and at the terminus of this segment on the said west line of Hagensburg to be elevation 236.4 feet.

(As an alternate, instead of referring the elevations to the center of the circle of the cylinder, the vertical control could be assigned to either the lowest point or the highest point of the cylinder.)

PARCEL 2:

A vertical shaft 6 feet square in and over that portion of the Rancho La Puente, in the County of ______________, State of ______________, as shown on a map recorded in book 1, pages 43 and 44 of Patents, in the Office of the County Recorder of said county, being 3 feet on each side of the following described center line and 3 feet westerly and easterly at right angles to said center line from the end point of said centerline:

Beginning at a point on the east line of said Rancho N 3° 22′ 36″ W 768.59 feet along said east line from its intersection with the northerly line of Foothill Road as described in deed to said county recorded November 3, 1972 in Book 1652, Page 38 of Deeds; thence S 72° 36′ 41″ W 420.00 feet to said end point. The elevation of the bottom of said shaft to be (based on USC & GS Station Los Cerritos, 356.2 feet,) elevation 290.0 feet and the top to be 8.0 feet above existing ground level.

PARCEL 3:

An easement for ingress and egress over a strip of land 20 feet wide over that portion of the Rancho La Puente, in the County of ______________, State of ______________, as shown on a map recorded in book 1, pages 43 and 44 of Patents, in the Office of the County Recorder of said County, the easterly line of which is described as follows:

Beginning at the intersection of the easterly line of said Rancho with the northerly line of Foothill Road as described in deed to said county recorded November 3, 1972 in Book 1652, Page 38 of Deeds; thence along said Rancho line N 3° 22′ 36″ W 800.00 feet, the northerly line to bear S 72° 36′ 41″ W from the northerly terminus thereof and the southerly line to be coincident with said northerly line of Foothill Road.

PARCEL 4:

A temporary easement for construction purposes in a strip of land 100.00 feet wide, in, over and under that portion of the Rancho La Puente, in the County of ______________, State of ______________, as shown on a map recorded in book 1, pages 43 and 44 of Patents, in the Office of the County Recorder of said County, the center line of which is described as follows:

Beginning at a point on the east line of said Rancho N 3° 22′ 36″ W 768.59 feet along said east line from its intersection with the northerly line of Foothill Road as described in deed to said county recorded November 3, 1972 in Book 1652, Page 38 of Deeds; thence S 72° 36′ 41″ W 1863.78 feet to a point on the west line of the land described in deed to William Hagensburg recorded February 9, 1972 in Book 1475, Page 24 of Deeds, said point being N 2° 46′ 12″ E 567.89 feet along said west line from the said northerly line of Foothill Road, the side lines of said easement to be shortened or lengthened to terminate at the above described east and west lines.

(The date or time period of termination of this temporary easement should be made a public record.)

In some cases, the top of subterranean rights can be established by the EXCEPTION form; i.e., after describing the planimetric shape, "EXCEPT that portion thereof lying above elevation 331.46 feet." If the stated level is ten feet below the surface of the ground, you have reserved that much for use by others.

Avigational

Since this type of easement has to do with airports which are under the direction of the Federal Aviation Agency (FAA), it would be well to consult with them relative to the establishment of any new avigational easements. There are FAA Technical Standards available from the U.S. Government Printing Office, Washington, D.C. and samples of a CLEAR ZONE (AVIGATION) EASEMENT and MODEL AIRPORT ZONING ORDINANCE are available from an FAA Area office.

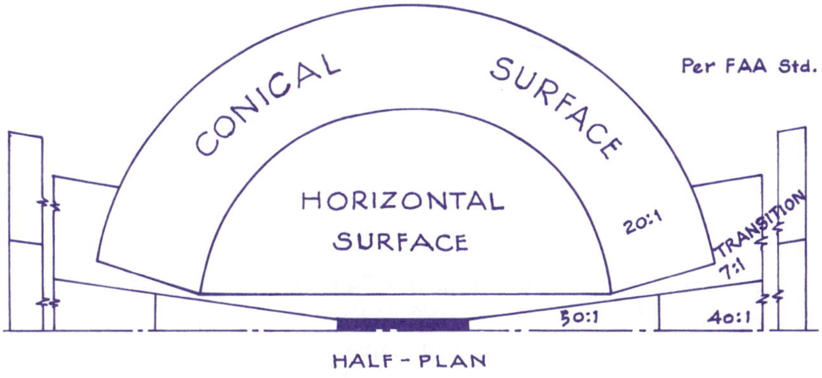

Fig. 106

In writing avigation easements, simplification is helped by separating the whole into parts. The vastness is not the problem but rather correlation of the graduated and conical planes into their proper juxtaposition. Start with the plane area of the runway itself and then fan out both laterally and vertically, or it might be expressed as diagonally aerially. Looking at the half-plan in Figure 106, the ratios of slope (50:1 etc.) are shown in the applicable areas outside of and beyond the runway.

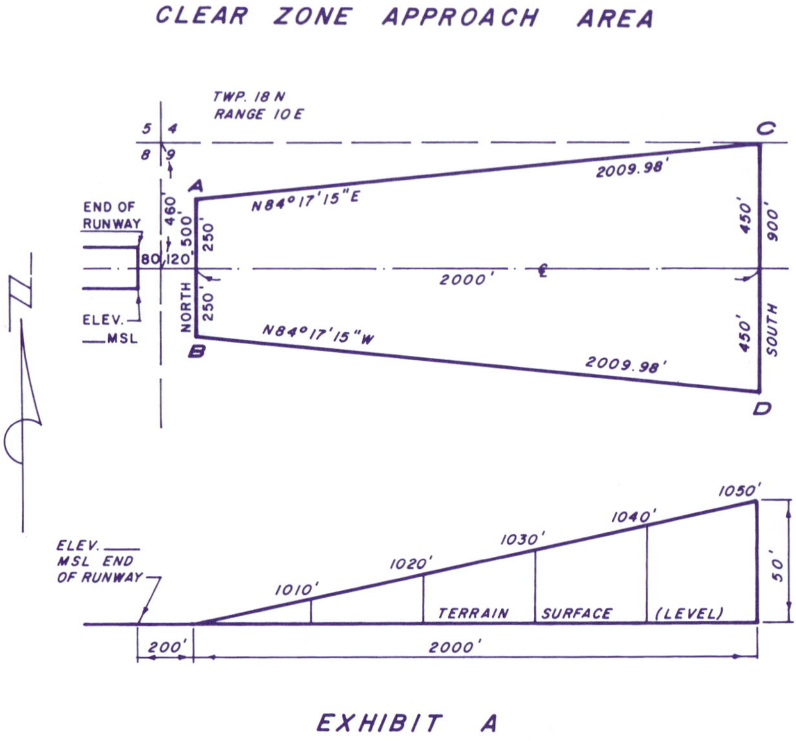

EXHIBIT A

Fig. 107

A typical plan and cross section of a CLEAR ZONE APPROACH AREA AT EAST END OF EAST-WEST RUNWAY (adapted from FAA Sample Form) is shown in Figure 107 and a suggested description for it reads thusly:

> A perpetual and assignable easement for a clear zone approach area over that portion of Section 9, T 18 N, R 10 East, of the ___ Meridian in the County of Nye, State of Nevada described as follows: Beginning at the intersection of the west line of said Section with the easterly prolongation of the center line of the east-west runway of Airville Municipal Airport, being southerly 460 feet from the north-west corner of said Section and on said prolongation East (bearing based on said center line per AMA Survey Plan No. 232-B) 80.00 feet from the end thereof; thence East 120.00 feet to the true point of beginning; thence North

> 250.00 feet; thence N 84° 17′ 15″ E 2009.98 feet; thence South 900.00 feet; thence N 84° 17′ 15″ W 2009.98 feet; thence North 250.00 feet to the true point of beginning. The easterly clear zone approach surface is a trapezoidal plane with a slope of 40 to 1 (one foot of elevation for each forty feet or horizontal distance) located directly above the easterly clear zone approach area, hereinabove described, which inclined plan has an elevation of __ feet above mean sea level at its inner and lower edge along line AB, as shown on Exhibit "A", and an elevation of __ feet above mean sea level at its outer and upper edge along line CD, as shown on said Exhibit "A."

Here too it would be desirable to include the drawing Exhibit "A," "attached hereto and made a part hereof."

To cover the entire area surrounding an airport with appropriate flight-pattern easements, there should be a complete set of plans showing not only topographic features and contours but also all of the aerial elevations in their respective areas; these would appear as contours in the air. Cross sections of the controlling surfaces in relation to the ground conditions are a most desirable adjunct to the plans. All of this would be further enhanced by the inclusion of a perspective drawing of at least one quarter of the entire layout, a simple example of which is shown in Figure 108.

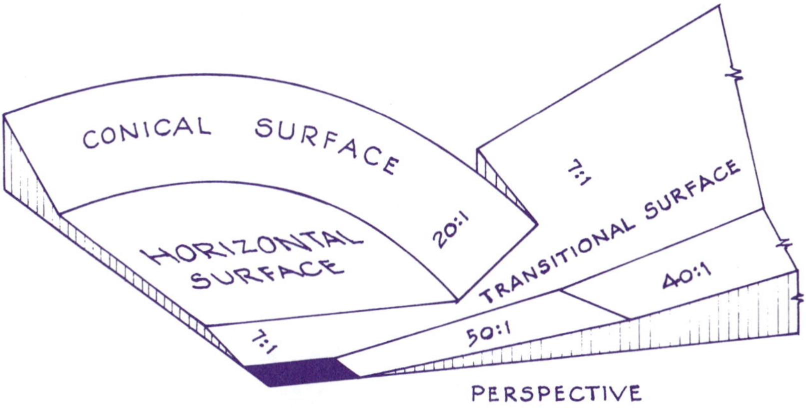

Fig. 108

The elevations-in-air are bottom limits developed from the physical conditions in and around the airport and are usually from an elevation up, that is to say, the sky is the limit on top with bottom planes set at elevations above sea level which leave a differential to the ground within which objects may exist without interference to aerial activities. A warning is in order, however, that if your client is planning to build any high-rise structure within these limited confines, be sure to check with the FAA regional office and get a clearance from them before starting construction. They can and do change their rules and being a governmental agency, they are not required to give public notice through the avenues of statutory constructive notice.

If the inclusion of such a set of plans in or with the recording of a deed is difficult because of their cumbersomeness, they could be filed in the official depository of public records and given a file number to which reference could be made in various documents pertaining thereto.

PROBLEMS

1. Cite an example of a description used on a deed of conveyance for a three-dimensional form which does not use metes and bounds.
2. What is the *fundamental* description used in either an aerial or subterranean description?
3. What is the *complementary* description used for either aerial or subterranean descriptions?

Appendix

Do's and Don't's

Here are a few reminders to watch in writing legal descriptions. The reasons behind these statements are to be found in the text.

* * * * * * *

Do be concise.

Do follow through.

Do be clear.

Do be positive.

Do check maps before using.

Do check legal meaning of questionable words.

Do determine whether survey or description controls.

Do test monuments for value and control.

Do review codes and court cases before making decisions.

Do watch relationship of bases of bearings.

Don't put distance figures next to lot numbers.

Don't omit *necessary* words and phrases.

Don't use unnecessary excess words and phrases.

Don't write into a questionable area before the problem is solved.

Don't be ambiguous.

Don't overburden an easement.

Don't use multiple incompatible ties.

Don't use references to quasi-public information in descriptions.

Do check veracity of information.

Do have sufficient sufficiency.

Do use correct references.

Do determine if your problem is subject to the laws of "public lands" or "private lands."

Do acknowledge all ties.

Do consider marketability of title.

Do question conflicting elements.

Do analyze mistakes.

Do capitalize cardinal directions of North, East, West and South.

Do capitalize the entire word of a cardinal direction if it is to be used as a bearing, i.e., NORTH.

Do end a caption with a colon when more is to follow.

Do use a semicolon at the end of each course except the last one.

Do consider surveyed lines.

Do respect one easement against another.

Do question "acquiescence" vs "agreement".

Do question "conveyed" vs "described".

Don't write "N'ly 15° E'ly or "northwardly 15° eastwardly."

Don't spell out numbers.

Don't capitalize a general direction such as "easterly."

Don't capitalize a direction used as an adjective such as "*south* line."

Don't create conflicting conditions.

Don't change descriptions that are already on record.

Don't cut off any title to land to which your client is entitled.

Don't give your client title to land to which he is not entitled.

Don't hurry to proportion.

* * * * * * *

DO NOT USE: —

EXCEPTIONS in the caption.

The word "either" in describing a strip of land as "____ feet either side of the following described line:."

"That part" when you have several portions.

"Those portions" when you have only one part.

Proration in a metes and bounds description.

"Adjacent" for describing the relationship of an easement to the parcel it serves.

The word "due."

"Front" and "rear."

List of Illustrations

Answers to Problems

Chapter 1

1. See page 1.1.
2. See page 1.2.
3. See page 1.2.
4. See page 1.3.
5. See page 1.4.
6. See pages 1.6 & 7.
7. See page 1.7.
8. See page 1.8.

Chapter 2

1. Assessor's Office and office of depository of public records.
2. County (or Parish) Surveyor's, City Engineer's (if in city), State Highway's, Railroad's, Sanitation District's, Recorder's and Public Utility's offices.
3. See page 2.4.
4. See pages 2.5 and 6.

Chapter 3

1. Yes. See page 3.2.
2. First paragraph on page 3.1.
3. See page 3.5.
4. See page 3.5.
5. See page 3.6.
6. No. See page 3.6.
7. No. See page 3.9.
8. No. See page 3.10.
9. No. See page 3.13
10. See pages 3.15 et seq.

11. Recommended way to write the easement:

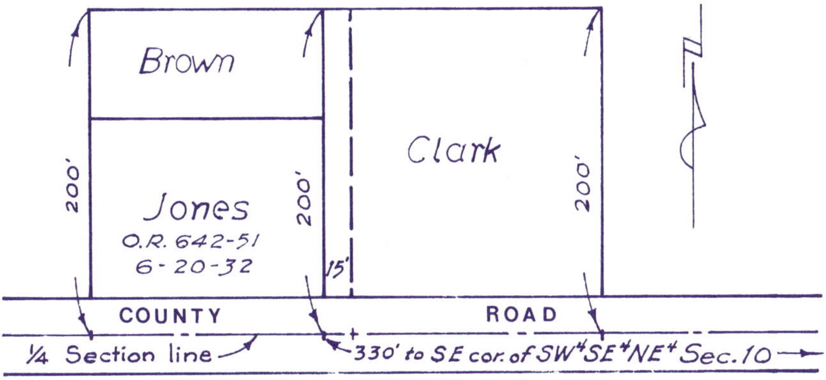

An easement for road and public utilities purposes over a strip of land 15 feet in width over a portion of the Southwest quarter of the Southeast quarter of the Northeast quarter of Section 10, Township 4 South, Range 10 East of the San Bernardino Meridian in the county of Riverside, state of California, according to an Official Plat of said land filed in the District Land Office, the west line of which is coincident with the east line of the land described in deed to R. L. Jones recorded June 20, 1932 in book 642, page 51, of Official Records, said east line being described as follows:

Beginning at a point on the center line of the County Road which is the south line of said Northeast quarter, 330.00 feet west of the Southeast corner of said Southwest quarter; thence true north 200.00 feet.

Subject to the interest in the County Road.

Chapter 4

1. No.
2. No.
3. See page 4.22.
4. Use even numbered element; see page 4.33.
5. See page 4.3.
6. See page 4.6.
7. See page 4.2.
8. See page 4.4.
9. See page 4.8.

Chapter 4 *cont'd.*

10. See page 4.11.

11.

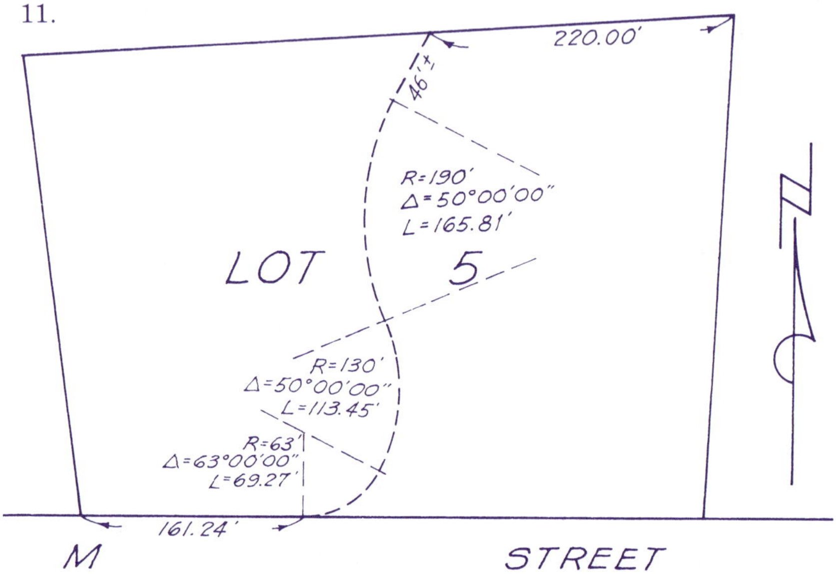

That portion of Lot 5, etc., etc. lying easterly of the following described line: Beginning at a point on the south line of said Lot N 89° 36′ E. 161.24 feet from the southwest corner thereof; thence on a curve tangent to said south line and concave to the north, having a radius of 63 feet, easterly and northeasterly 69.27 feet along said curve through a central angle of 63° to the beginning of a compound curve concave to the west, having a radius of 130 feet; thence northerly 113.45 feet along said curve through a central angle of 50° to the beginning of a reverse curve concave to the east, having a radius of 190 feet; thence northerly 165.8 feet more or less to a tangent line which passes through a point on the northerly line of said Lot distant westerly 220.00 feet feet along said line from the northeast corner thereof; thence northerly 46 feet more or less to said north line.

Comment: Note that the first curve was necessarily cited as tangent to the lot line because no prior course or statement had established the relationship. Also note that the first two curve descriptions include the central angle but the third one did not and furthermore, the distance was flagged "more or less" in order to allow it to give to the requirement of a *tangent* line meeting the specified point on the north lot line.

12. See completed drawing.

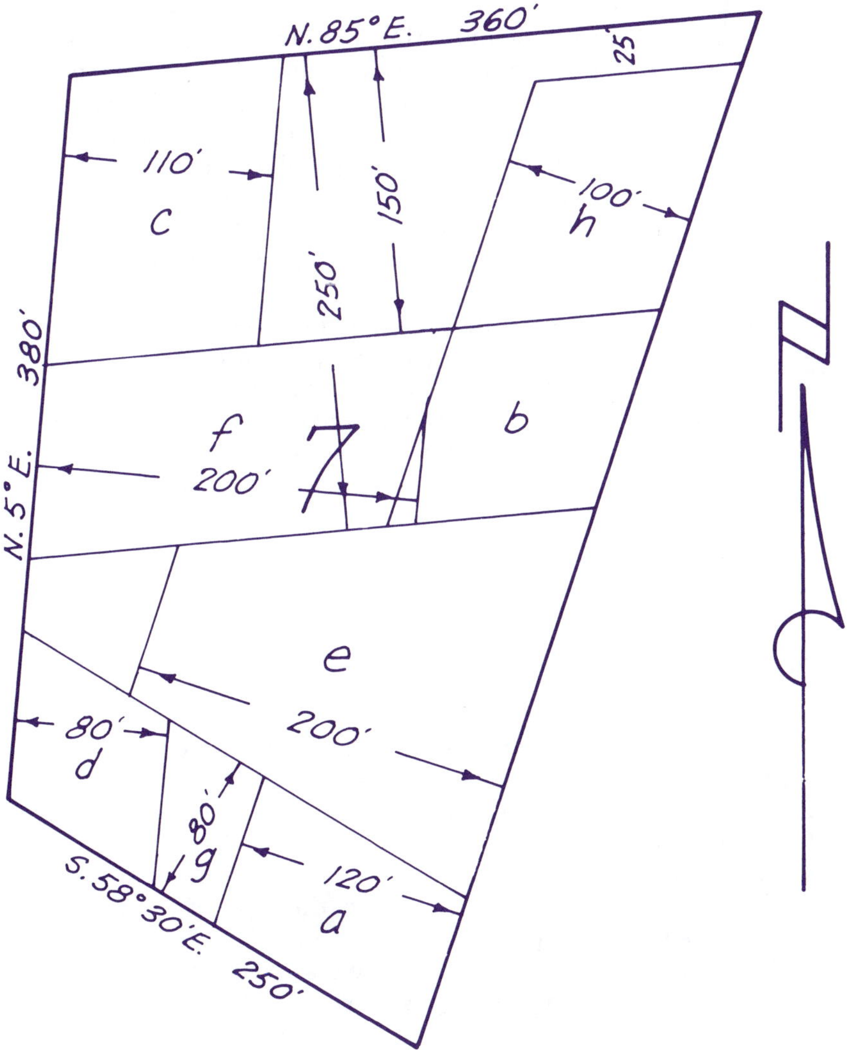

Chapter 4 *cont'd.*

13. A simple solution with just one document would be for "X" to quitclaim to "A," "B" and "C" as their respective interests appear of record:

    All those portions of Lot 17 not included within the following described land:

    > Beginning at a point on the north line of said Lot West 60 feet from the northeast corner thereof; thence along said north line West 37 feet to a line that is parallel with and easterly 50 feet measured at right angles from the west line of said Lot; thence along said parallel line southerly 117 feet to the westerly prolongation of the south line of Lot 18; thence along said prolongation East 68 feet to a line which bears South from the point of beginning; thence North to the point of beginning.

    Another way to resolve such a problem, especially if there were ambiguities in several of the parcels, would be to convey each parcel *as originally recorded* to a Trustee who would then use correct descriptions on new documents and convey the respective areas to the corresponding owners in a proper manner.
    Either way would remove any cloud of title imposed on the lands of "A," "B" and "C" by the original grant deeds.
    The words "more or less" might be used after the distances, "37", "117" and "68" but the ties will hold regardless.

Chapter 5

1. Not without words of conveyance in the agreement or in separate documents.
2. When recorded; see page 5.11.
3. "agreement line."
4. No; see page 5.4.
5. One by reference to "track" location, the other by reference to a described line; see page 5.18.
6. Characteristics, inherent or applied; see page 5.12.
7. Depends on form of wording; see page 5.7.
8. No; see page 5.7.
9. Intent.

Chapter 6

1. MHW and MLW page 6.5.
2. 18.6 years; page 6.5.
3. No; page 6.6.
4. No.
5. In whatever manner the ownership existed prior to creation of the man-made body of water; page 6.15.
6. See page 6.18.
7. Sometimes but not always.

Chapter 7

1. Yes.
2. Yes, because the grantor cannot sell what he does not own.
3. Yes.
4. See page 7.10.
5. The value is based on the chronological relationship. Page 7.13.
6. Acquires only part owned by subdivider; page 7.19.
7. See page 7.18.
8. No. See page 7.20.
9. No. See page 7.22.
10. Public lands and private lands. See page 7.22.
11. Each half must contain an area equal to the other half; page 7.23.

Chapter 8

1. See pages 8.1 and 2.
2. See page 8.2.
3. Only if adjoining deed description is OK. See page 8.3.
4. Not if road deed is fee or has some other qualification such as an excepted reversionary right.
5. Underlying interest. See page 8.3.
6. No. See Earl v. Dutour and consider custom in an area per Ferris v. Simmons. See page 8.3.
7. Yes.
8. Yes.
9. Yes. See page 8.5.

Chapter 8 *cont'd.*

10. No. On Public Land Surveys — yes but otherwise not necessarily. See page 8.8.
11. No. See pages 8.9 and 10.
12. Yes. See page 8.10.
13. See page 8.9.
14. No. See page 8.10.
15. Background, character and recognition. See page 8.7.
16. See page 8.16.

Chapter 9

1. No. See page 9.2.
2. "to (w) line of _____ Street; thence along (w) line of _____ Street." See page 9.3.
3. See pages 9.4 and 5.
4. Usually an easement status but it depends on wording and/or prior title. See page 9.5.
5. No. See page 9.10.
6. No. See page 9.10.
7. See page 9.7.
8. See page 9.9.
9. No. See page 9.10.
10. Center line, survey line and offset line. See page 9.10.
11. Description of part of Parcel:

    That portion of Parcel 1 in the City of La Habra, county of Orange, state of California, as per map recorded in book 1, page 40, of Parcel Maps in the office of the County Recorder of said county lying easterly of the centerline of the 10 foot Easement for Sanitary Sewer to said city of La Habra as shown on said Parcel Map.

    The following is a recommended form of *metes and bounds* description for the requested problem.

    That portion of the Northeast Quarter of Section 6, Township 3 South, Range 10 West in the Rancho La Habra, city of La Habra, county of Orange, state of California

PARCEL MAP

Portion of NE ¼ of Section 6,
T3S, R10W, Ro. La Habra,
M.M. 51/7,
La Habra City, Calif.

1943
FEB 10 1966
FILED FOR RECORD THIS ____ DAY OF ________
AT ____ M. IN BOOK 1 PAGE 40
OF PARCEL MAPS COUNTY OF ORANGE, CALIFORNIA
AT REQUEST OF COUNTY SURVEYOR
$5 00 J. WYLIE CARLYLE, County Recorder

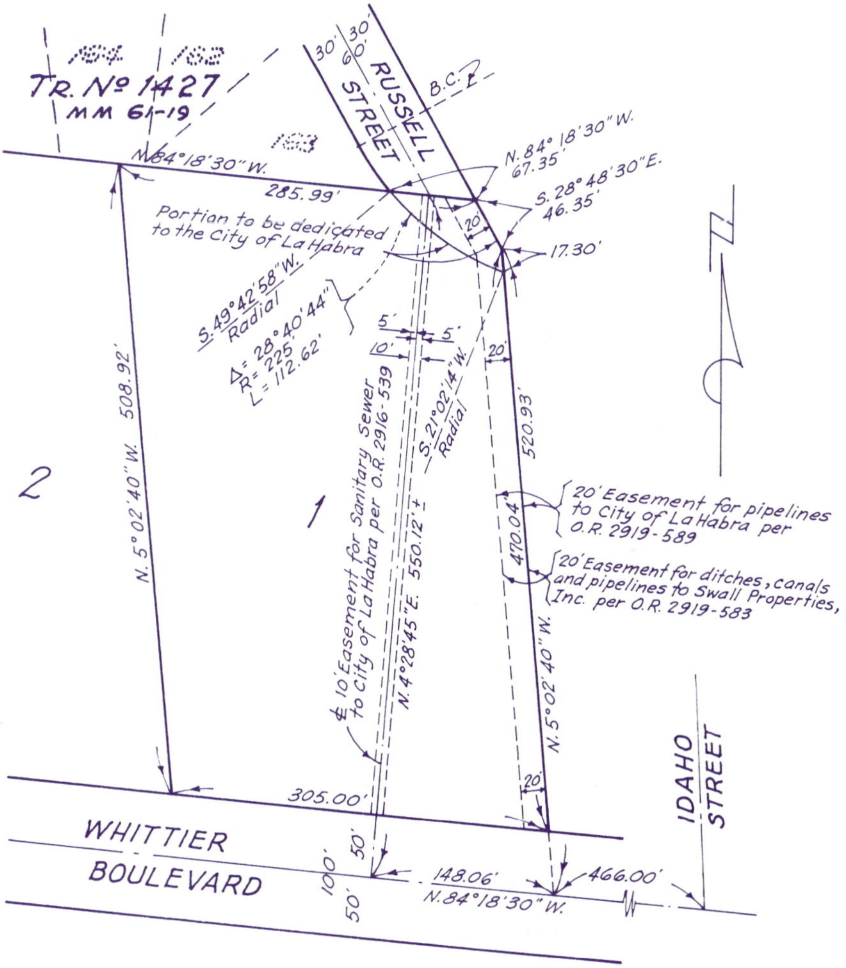

Chapter 9 *cont'd*

as per map recorded in book 51, page 7, of Miscellaneous Maps in the office of the County Recorder of said county, described as follows:

Beginning at a 2″ IP with tag LS 2489 at the southeast corner of Tract No. 1427 as per map recorded in book 61, page 19, of said Miscellaneous Maps, and the most northeasterly corner of Parcel 1 shown on a map recorded in book 1, page 40, of Parcel Maps in the said office of the County Recorder; thence along the southeasterly prolongation of the northeasterly line of RUSSELL STREET as shown on said maps South 28° 48′ 30″ East 46.35 feet to a 2″ IP with tag LS 2489; thence South 5° 02′ 40 East 470.04 feet along the easterly line of said Parcel 1 to WHITTIER BOULEVARD shown on said map; thence westerly along WHITTIER BOULEVARD to the center line of the 10 foot Easement for Sanitary Sewer to said city of La Habra as shown on said Parcel Map; thence northerly along said center line to the south line of said Tract No. 1427; thence easterly to the point of beginning.

(Note the form that carries title to any portion in WHITTIER BOULEVARD which the owner of the land adjoining may have.)

Chapter 10

1. Sometimes. See page 10.4.
2. No. See page 10.3.
3. Mistake. See page 10.4.
4. See page 10.8.
5. See page 10.5.
6. See page 10.9.
7. See page 10.7.

Answer to ASSIGNMENT

That portion of the SW ¼ of Section 2, T 5 S, R 3 W, of the 9th Meridian North in the County of Mena, State of North Dakota according to an Official Plat of said land filed in the District Land Office described as follows: Beginning at the intersection of the west line of the R/W described in deed to the Intermountain Power Co. recorded in Book 201, Page 72, of Deeds, records of said county, with the center line of State Highway #201 as described in Book 888, Page 32, of Deeds; thence along said center line S 78° 12′ 35″ W 148.72 feet to the beginning of a curve described therein concave to the south and having a radius of 440 feet; thence westerly 162.95 feet along said curve to the northerly line of the C.P.R.R. Co. R/W as described in deed recorded in Book 213, Page 97, of Deeds; thence N 68° 06′ W 224.73 feet to the southeast corner of the land described in deed to E. L. Jones recorded in Book 691, Page 36, of Deeds, being S 68° 06′ E 57.00 feet from the easterly line of "A" Street as shown on the map of SNOW TRACT recorded in Book 2, Page 31, of Maps of said County; thence along the easterly line of said Jones' land N 38° 14′ E 161.00 feet and N 12° 23′ W 53.21 feet to the northeast corner thereof which is on the north line of said SW ¼ distant along said line easterly 150.00 feet from the easterly line of said "A" Street; thence along said north line N 89° 41′ 05″ E 399.98 feet to said west line of the Intermountain Power Co. R/W; thence S 5° 13′ E 172.96 feet to the point of beginning.

SUBJECT TO:

a — Valley View Flood Control Dist. R/W over the southeasterly portion of said land as per deed recorded in Book 3416, Page 213, of Deeds.

b — Rights of the public in State Highway #201 over the southerly 40 feet of said land as per deed recorded in Book 888, Page 32, of Deeds.

c — The use of a triangular portion of said land approximately 18 feet by 50 feet in the northwest corner by the occupants of the land adjoining on the west by reason of a fence encroaching onto the subject land.

NOTE:

By reason of fences encroaching onto adjoining property, users of subject property also make use of approximately 18 feet of the Intermountain Power Co. R/W on the east and of a triangular portion in the southwest portion to a maximum of 1.8 feet.

Chapter 11

1. See page 11.3.
2. See page 11.2.
3. Yes. See page 11.5.
4. No. See page 11.7.
5. No. See page 11.7.
6. See page 11.9.
7. See page 11.11.
8. Any number.
9. No. See page 11.12.
10. Statutory, actual or constructive. See pages 11.20 et seq.
11. By referencing the map into a document which by due process establishes that document as an instrument of constructive notice. See page 11.22.
12. No. See page 11.23.
13. See page 11.23.
14. No. See page 11.24.
15. Go in the same direction on lines that are common to adjoining parcels. See page 11.26.
16. See drawing for Problem 16.
17. See drawing for Problem 17.

    This problem is to alert you to read *and* draft into the picture the boundary deeds first and then hold the ties.

    Notes:

    1 — The first tie "to a point on the south line" of Vandevort's land produces a different direction than that cited.

    2 — The tie to the "one inch iron pipe" must hold.

    3 — The tie "to the easterly line of . . . . . . Gould" does not specify any relationship to his southeast corner.

    4 — The bearing of "N 30° 20′ 00″ W" does not agree with position of Gould's easterly line by its tie of "395.00 feet from the southwest corner."

    5 — Do not forget that the street and lot are separate and that measurements in the lot are from the side line of the street — NOT the center line.

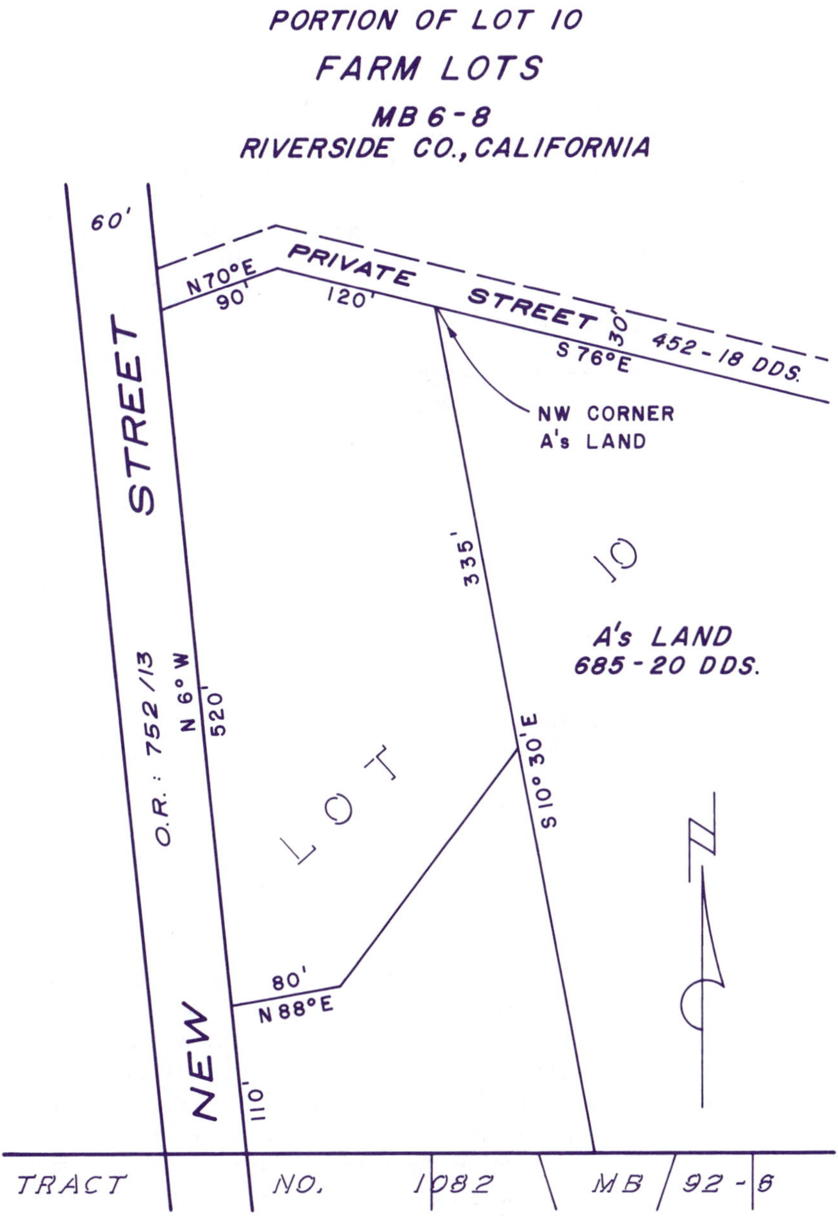

Answer to Problem 16

Chapter 11 *cont'd*

6 — Is the reservation of "an easement . . . . . over the north 30.00 feet of the south 435.45 feet of said lot" meant to apply to only Parcel 2 or to both Parcels 1 and 2? This matter of tacking on a reservation without specifying whether it applies to one particular parcel or all of them appears every once in awhile; you can see the question it creates and it is a good example of why it is necessary to be explicit.

18. Parcel 1: The southerly 100.00 feet of the westerly 100.00 feet, measured along the westerly and southerly lines respectively, of Lot 19 in Tract ___ in the County of ________ ____________, State of ____________________________, as per map recorded in Book ___, Page ___, of Maps in the office of the County Recorder of said County.

Parcel 2: A 20 foot easement for road purposes over a portion of Lot 19 in Tract ___ in the County of ________________, State of __________________________, as per map recorded in Book ___, Page ___, of Maps in the office of the County Recorder of said County, the southeasterly line of which passes through the northeast corner of said lot and the northwesterly line of which passes through the northeast corner of the southerly 100.00 feet of the westerly 100.00 feet, measured along the westerly and southerly lines respectively of said Lot.

Parcel 3: That portion of Lot 19 in Tract ___ in the County of ________________, State of ____________________, as per map recorded in Book ___, Page ___, of Maps in the office of the County Recorder of said County, lying southeasterly of a line which passes through the northeast corner of the southerly 100.00 feet of the westerly 100.00 feet, measured along the westerly and southerly lines respectively of said Lot and is 20.00 feet measured at right angles northwesterly from the northeast corner of said Lot. EXCEPT any portion thereof included within the westerly 100.00 feet measured along the southerly line of said Lot. RESERVING an easement for road purposes over the northwesterly 20.00 feet thereof.

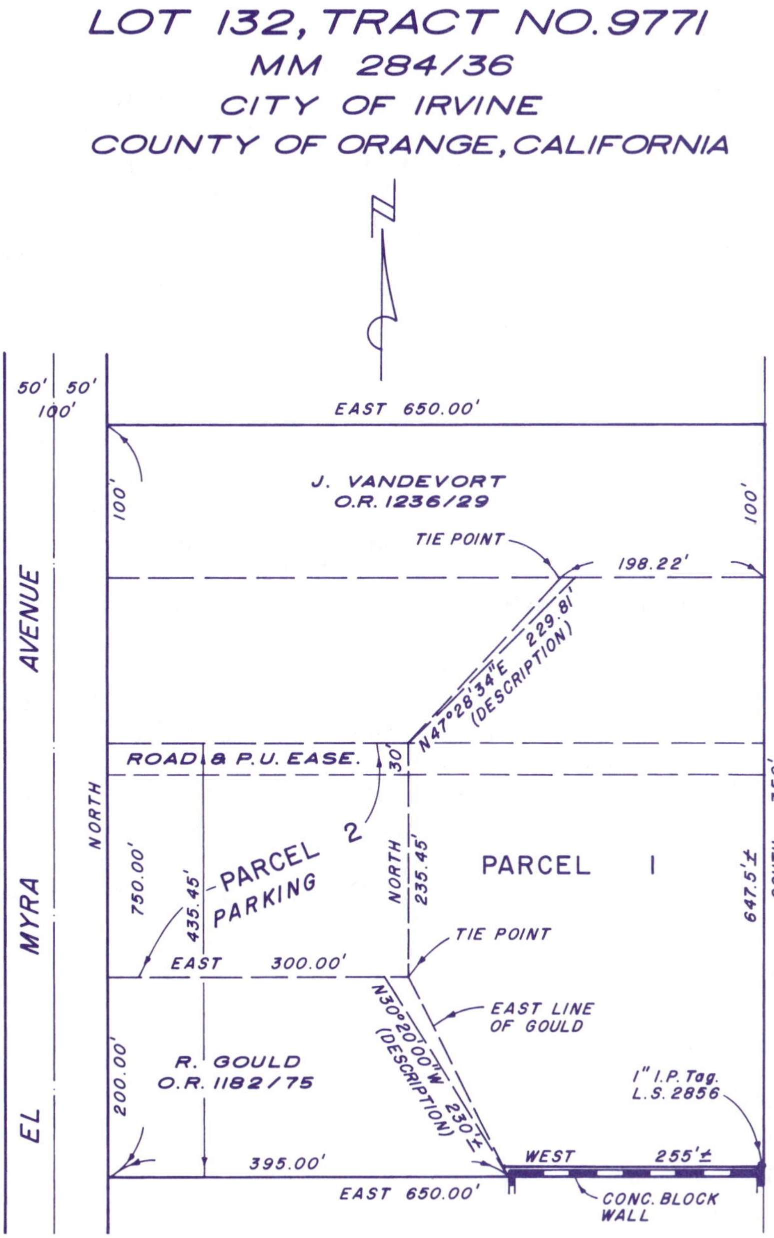

Answer to Problem 17

Chapter **11** *cont'd*

19. A suggested answer to Problem 19 reads as follows:

    That portion of the North one half of the North one half of the Northeast quarter of the Southeast quarter of the Southeast quarter of Section 17, Township 5 South, Range 10 West, in the Rancho Las Bolsas, County of Orange, State of California, as shown on a map recorded in Book 51, page 12 of Miscellaneous Maps, in the office of the County Recorder of said county, described as follows:

    Beginning at the Southeast corner of Tract 4957, as shown on a map recorded in Book 177, pages 23 and 24, Miscellaneous Maps, records of said County, said corner being on the center line of Euclid Street; thence South 88° 09′ 20″ West 368.36 feet along the South line of said Tract 4957, being also the South line of the North 2 acres of the East one half of the Southeast quarter of the Southeast quarter of said Section 17; thence South 1° 50′ 40″ East 65.85 feet; thence North 88° 09′ 03″ East 367.04 feet to the center line of said Euclid Street; thence North 0° 41′ 40″ West 65.86 feet to the point of beginning.

    EXCEPT that portion thereof, if any, included within the South 1 acre of the north one half of the north one half of the northeast quarter of the southeast quarter of the southeast quarter of said Section 17.

    Note: The lack of a tie following the distance of "368.36 feet" introduces the question of whether that is the distance to the west line of the NE ¼ NE ¼ SE ¼ SE ¼ of the Section or not.

Chapter **12**

1. See page 12.3.
2. See page 12.5.
3. See page 12.10.
4. See page 12.2
5. Any number.
6. Parcel U-1: Lot ___ in Tract No. ___ in the City of ________ ________________, County of ________________, State of ________________, as per map recorded in Book

___, Page ___, of Maps in the office of the County Recorder of said County.
EXCEPT the easterly 400.00 feet thereof.
ALSO EXCEPT the southerly 397.00 feet thereof.

Parcel U-2: An easement for road purposes over the easterly 400.00 feet of the northerly 25.00 feet of the southerly 422.00 feet and over the easterly 25.00 feet of the southerly 397.00 feet of Lot ___ in Tract No. ___ in the City of ______________, County of ______________, State of ______________, as per map recorded in Book ___, Page ___, of Maps in the office of the County Recorder of said County.

Parcel V-1: The easterly 400.00 feet of Lot ___, in Tract No. ___ in the City of ______________, County of ______________, State of ______________, as per map recorded in Book ___, Page ___, of Maps in the office of the County Recorder of said County.
EXCEPT the southerly 397.00 feet thereof.
RESERVING an easement for road purposes over the southerly 25.00 feet thereof.

Parcel V-2: An easement for road purposes over the easterly 25.00 feet of the southerly 397.00 feet of Lot ___ in Tract No. ___ in the City of ______________, County of ______________, State of ______________, as per map recorded in Book ___, Page ___, of Maps in the office of the County Recorder of said County.

7. Give the *direction* of the line to which stationing is being assigned. See page 12.14.
8. See pages 12.16 and 17.

Chapter **13**

1. No. See page 13.7.
2. Usually by reference to the acreage assigned to the lot as shown on the official plat. See page 13.9.
3. No. See page 13.10.
4. Not necessarily. See page 13.12.
5. The map made by the private surveyor, whether it is recorded or not. See page 13.13.

Chapter 14

1. The most common is by reference to a UNIT as a part of a condominium project.
2. The fundamental description is of the planimetric form. See page 14.7.
3. The complementary description is the control of the vertical limits. See page 14.7.

The Author

From a collocation of experience in land surveying, civil engineering, title engineering, architecture, planning and construction covering a span of forty-five years, Gurdon Wattles has had an association with and first-hand knowledge of the many facets of the subject matter upon which he discourses.

Besides teaching private and collegiate classes in survey boundary control and legal descriptions, the author has conducted many seminars in these and related subjects at the request of professional, governmental and private industry groups. In balance, he serves as a member of Surveying Technology Advisory Committees to a number of colleges.

Many of his articles have been published in national magazines and other media, besides which he holds copyrights on maps, books and brochures.

Mr. Wattles, a Fellow in ASCE, has been a member of the Committee on Land Surveying in the Division of Surveying and Mapping of the American Society of Civil Engineers for seven years and was its National Chairman 1969-1970; he was also National Chairman of the Committee for Improvement of Land Title Records of the American Congress on Surveying and Mapping, 1970-1972, and a member of that Committee seven years; he has been a member of the American Right of Way Association twenty years; as a member of the Planning Committee of the American Bar Foundation's preparation for a conference on Compatible Land Identifiers, he subsequently participated in its 1972 Conference. Mr. Wattles is also a member of the recently-created League of California Surveying Organizations, Past President of the Orange County Engineering Council and First Past Chairman of the Surveying and Mapping Technical Group of the Los Angeles Section, American Society of Civil Engineers.

The author recently retired from his position as Assistant Vice President and Consulting Title Engineer of Title Insurance and Trust Company and Pioneer National Title Insurance Company. He is a Registered Land Surveyor in California and Nevada. He is listed in "Who's Who" and reported in "Men of Achievement" international.

Index

Instructions for Survey of Public Lands — 1.3, 13.1, 13.4